The Parameter Space Investigation Method Toolkit

The Parameter Space Investigation Method Toolkit

Roman Statnikov

Alexander Statnikov

ARTECH HOUSE

BOSTON | LONDON

artechhouse.com

Library of Congress Cataloging-in-Publication Data
A catalog record for this book is available from the U.S. Library of Congress.

British Library Cataloguing in Publication Data
A catalogue record for this book is available from the British Library.

Cover design by Merle Uuesoo

ISBN 13: 978-1-60807-186-9

10 9 8 7 6 5 4 3 2 1

Contents

Acknowledgments

We authors would like to thank David Olson and Ralph Steuer for their interest in our work. We are also grateful to Dan Boger, Alex Bordetsky, Eugene Bourakov, Vladimir Dobrokhodov, Roberto Cristi, Isaac Kaminer, Fotis Papoulias, and Terry E. Smith (all from the Naval Postgraduate School); Petr Ekel (University of Minas Gerais, Brazil); Enric Xargay (University of Illinois at Urbana-Champaign); Mohamed E. Elmadawy and Mohamed A. El Zareef (Mansoura University, Faculty of Engineering, Structural Engineering Department, Egypt); Kivanc Ali Anil (Turkish Navy); Vladimir Astashev, Rivner Ganiev, Josef Matusov, Il'ya Sobol', and Konstantin Frolov (Russian Academy of Sciences) for their help and contribution. We are especially thankful to Irina Statnikova for her ongoing help and support.

The manuscript of this book was the basis for the lecture course "Multicriteria Analysis" created by R. Statnikov at the Naval Postgraduate School in Monterey, California. We would also like to thank Raymond (Chip) Franck, who supported this lecture course, and Tom Hazard and Wally Owen, who promoted this course.

Preface

Real-Life Optimization Problems and Parameter Space Investigation Method

What Is This Book About?

While searching for optimal solutions, two questions arise: where to search and how to search. Various optimization methods, to which an uncountable set of works is devoted, answer the second question. However, without correctly answering the first question, the search for optimal solutions can lead to unsatisfactory results. Usually such "optimization" is observed when solving real-life problems. Notice that, with the rare exception, very little attention is paid to a fundamental problem such as the determination of the feasible solution set or where to search for optimal solutions. Defining this set is directly related to the correct statement of the optimization problem. This usually represents the most significant difficulties for the expert. Therefore, in the majority of cases, the expert cannot state the problem correctly and ends up solving ill-posed problems. In this book, we will show how to define the feasible solution set and thus answer the fundamental question of where to search for optimal solutions.

Difficulties in Stating an Optimization Problem

The majority of engineering problems (design, identification, design of controlled systems, large-scale systems, predicting from observational data, and so on) are essentially multicriteria. These problems are encountered in all facets of human activity. The expert always wishes to optimize not one important criterion, but all of the most important criteria, many of which are antagonistic. The basic components of a problem statement are constraints on criteria, design variables, and so-called functional dependences. These constraints determine the feasible solution set, a region in the criteria and design variable spaces where optimal solutions should be sought. The determination of the feasible solution set is the essence of problem statement. An important subset of the feasible solution set is referred to as the Pareto optimal solution set. A solution[1] is Pareto optimal if value of one criterion can be improved only at the expense of worsening at least one of the other criteria. In order to solve the optimization problem, one has to identify the Pareto optimal set. Obviously, if the feasible solution set has been determined incorrectly or incompletely, the obtained Pareto solutions may not have practical value. Nowadays this situation is quite typical in the overwhelmingly majority of engineering problems.

Nature of Constraints

Performance criteria (goal functions) (e.g., the fuel consumption, cost, efficiency, and so on) should be optimized. It is desired that, with other things being equal, these criteria would have the extremal (e.g., maximum or minimum) values. Since criteria are contradictory, the definition of criteria constraints represents significant, sometimes insurmountable difficulties [1–5]. Furthermore, there are functional dependences. Unlike criteria, functional dependences do not need to be optimized. It is required that only their respective constraints are satisfied. We recognize two kinds of functional constraints: rigid and "soft" (nonrigid). For example, standards are rigid functional constraints. These constraints are not supposed to be changed—they are known a priori. On the other hand, "soft" functional constraints (e.g., overall dimensions) can be changed. Quite often the correct definition of these constraints is also difficult for the expert. If functional constraints are poorly defined, many interesting solutions become unreasonably unfeasible. As a result, the feasible solution set can be empty.

Functional dependences and criteria depend on design variables (e.g., geometric sizes). Design variables are changed within some boundaries. Quite often these boundaries can be revised, if it leads to the improvement of values of the main criteria.

Notice that in real-life problems, the number of functional and criteria constraints can reach many dozens, if not hundreds, and the dimensionality of design variable vector can reach many hundreds and thousands.

In the traditional statement of optimization problems, constraints are usually given a priori. However, it is unlikely that such constraints are correct, especially given the high dimensionality of the problems and the complexity of

1. Vilfredo Pareto (1848–1923) was an Italian economist. A strong definition of the Pareto optimal set is given in Section 1.1. From now on, we will be using the expression "Compromise solutions" to refer to "Pareto optimal solutions."

the mathematical model. That is why it is necessary to ensure the correctness of given constraints. Otherwise, the optimization can lead to the meaningless results or equally to the loss of important solutions. As mentioned earlier, established optimization methods do not address the problem of defining the feasible solution set.

Taking into account the difficulties of determining constraints, the feasible solution set can be poor or even empty. Therefore, it is very important to help the expert determine the constraints correctly.

Now, let us turn to the example that illustrates the main topic of this book. These days it is impossible to imagine a doctor, even the most gifted one, working without diagnostic tools, such as X-ray, tomography, lab tests, and so on. Likewise, in engineering problems, it is difficult to imagine approaching the challenging tasks without the tools for constructing and analyzing the feasible solution set. The tools that an expert should use to state and solve real-life problems are discussed in our book.

The Parameter Space Investigation (PSI) Method

In order to construct the feasible solution set, a method called the PSI method has been created and successfully integrated into various fields of industry, science, and technology [1–5]. This method has been used in designing the space shuttle [3, 6–8], nuclear reactors [3], unmanned vehicles, [3, 9], aviation [3, 10–13], cars [3, 14–17], pumping units [18], ships [2, 19–22], metal tools [2, 5, 23], bridges [24], wind power system [25, 26], wireless battlefield networks [27], energy efficient sensor networks [28], and robots [29]. The PSI method is based on the systematic investigation of the multidimensional domain [1–5, 30–36]. A computer generates multidimensional points (each point corresponds to a certain design). This is accomplished by uniformly distributed sequences, nets, and quasi-random points. Then the computer defines the values of criteria in these points. In a continuing dialogue between an expert and a computer, the constraints are repeatedly revised, and, as a result, the feasible and Pareto optimal solutions are determined. Thus, an expert can assess the price of making concessions in various constraints (i.e., what are the losses and the gains). Prior experience has shown that the expert is often ready to change constraints by having information on a sufficient improvement of the values of the main criteria. An expert obtains such information on the basis of the PSI method.

In the PSI method, stating and solving problems is a single process. Such an interactive mode allows us to take into account the experience and knowledge of the expert. As a rule, using the PSI method leads to the correction of the initial problem statement, including the correction of constraints and the mathematical model. From our point of view, all real-life optimization problems have to be stated and solved in an interactive mode.

Sometimes using the PSI method can demand carrying out computationally expensive experiments. In these cases most of computer time is spent on determining the feasible solution set, the correction of the problem statement that eventually leads to obtaining the justified optimal solutions, and there is no way around it.

To the best of our knowledge, the PSI method is the only available method for solving the fundamental problem of constructing the feasible solution set.

The PSI method is implemented in the MOVI (Multicriteria Optimization and Vector Identification) software system [37]. The PSI method and the MOVI software system can be universally applied to many problems and only require access to the mathematical model of the system or object under consideration. Even when a model is not available, the PSI method can still be applied to an approximate mathematical model that can be derived from observational data using statistical machine learning classification and regression algorithms [38–42].

The PSI method and MOVI system provide tools for constructing and analyzing the feasible solution set. First of all, these are the test tables. Other tools are tables of feasible and Pareto optimal solutions, histograms, tables of functional failures, and graphs of criterion versus design variable and criterion versus criterion. All of these tools provide us with unique information about: (1) the distribution of feasible solutions in the design variable and criteria spaces, (2) the work of all constraints, (3) the expediency of their modification, and (4) resources for improvement of the object.

The Number of Criteria

Consider the important issue of the number of criteria in a real-life problem. This number must be no less than necessary. The greater the number of criteria taken into account, the greater the information obtained about: (1) the resources of improving the object (ship, car, nuclear reactor, aircraft, machine tool, robot, submarine), (2) the performance of a mathematical model and constraints, and (3) the accuracy with which the criteria are calculated and how much one can trust them.

The PSI method allows us to consider as many criteria as necessary. For example, in the problems of vector identification, the number of criteria reaches many dozens [5, 28, 29].

Using Single-Criterion Methods to Solve Real-Life Problems

In the overwhelming majority of cases, attempts are made to present real-life multicriteria problems as single-criterion problems. In this case, the expert optimizes only one criterion and imposes constraints on other criteria. This approach is appealing because of the apparent simplicity of solving a complex problem. However, there is something else more important that calls in question the competence of this decision.

Above all, the complex problem of determining the feasible solution set is shifted onto the expert's shoulders. Unfortunately, the expert is usually unable to do this. As a result of substituting a single-criterion problem for a real-life one, we end up with a problem that has little to do with real life. Therefore, the numerous attempts to reduce a multicriteria problem to a single-criterion problem result in "throwing out the baby with the bathwater." The search for optimal solutions without determining the feasible solution set is substituting myth for reality. In other words, there are two alternatives: do it simply, or do it right.

Using single-criterion methods without substantiation of the feasible solution set does not guarantee that the obtained optimal solutions are feasible ones. Furthermore, the expert does not have information about compromise solutions considering all criteria.

Multicriteria Identification

Usually in optimization problems we assume by default that the adequacy of the mathematical model is beyond question (i.e., performance criteria adequately describe the investigated object). However, in the majority of cases it is not true. For this reason multicriteria identification of mathematical models is of fundamental importance in real-life problems. The central point is the construction of the feasible solution set in multicriteria identification problems. These problems are encountered in the production of machine tools, automobiles, ships, and aircraft, where enormous amounts of money are spent on operational development of a prototype. Questions of the multicriteria identification and adequacy of the mathematical model will be discussed in our book.

After the PSI method was developed and used successfully, it became necessary to write a new book where we synthesized the extensive experience of applying the PSI method to a variety of engineering problems. We used it as the basis for a lecture course "Multicriteria Analysis," which is taught in the United States and Russia. The material of the book is set out in a popular, concise form, with a large number of illustrations. The book is intended for a wide circle of readers, from undergraduate to graduate students, to researchers and experts involved in solving applied optimization problems. Reading this book does not require any special mathematical education.

This book is organized as follows. Part I gives an overview of the PSI method and MOVI software system. Specifically, Chapter 1 provides an introduction to multicriteria analysis. Chapter 2 discusses the PSI method that allows us to construct the feasible and Pareto sets and provides an example of a typical engineering optimization problem. In Chapters 3 and 4 the visualization tools for multicriteria analysis are demonstrated by means of this example where the main possibilities of the MOVI software are illustrated. We show the most general case, when an expert requires help for the definition of the feasible solution set.

In Part II, we describe the statement and solution of real-life optimization problems[2]. The choice of real-life examples is motivated by the following considerations: it is desirable to briefly show the process of searching for optimal solutions in real problems without going into details of describing the models. Part II has the following organization:

- In Chapter 5 we demonstrate examples of improving prototypes of a ship, a truck frame, a rear axle housing, and an orthotropic bridge. The application of the PSI method in combination with the finite element analysis is discussed in Sections 5.3 through 5.5.
- As already noted, one of the major aspects in engineering optimization is the adequacy of the mathematical model to the actual object. This problem is discussed in Chapter 6. In some cases, in order to identify one object, it is necessary to work with several models simultaneously (see Section 6.3).
- In Chapter 7 we discuss other multicriteria problems and related issues. We often face an important problem where the desired solution is unattainable. The search for the compromise solution for parametric optimization problem is shown in Section 7.1. The problem of the design of controlled engineering systems is presented in Section 7.2. In many cases there are no a priori mathematical models. However, there are available observations in the form of tables that give an indication of the behavior of the system under investigation. Multicriteria analysis from observational data is considered in Section 7.3. For many applied optimization problems, it is necessary to carry out a large-scale numerical experiment in order to construct the feasible set. Multicriteria optimization of large-scale systems in the parallel mode is illustrated in Section 7.4. We also discuss the number of trials necessary for the statement and the solution of the real-life problems in Section 7.5.
- Chapter 8 considers the new paradigm of database search [43]. The advent of the World Wide Web made search engines the most essential component of our everyday life. However, the analysis of information

2. If the descriptions of experiments presented in some chapters of Part II are not relevant for some readers, you can skip the corresponding chapters.

provided by current search engines often presents a significant challenge to the client. This is, to a large extent, because the client has to deal with many alternatives (solutions) described by contradictory criteria when selecting the most preferable (optimal) solutions. In such situations, the construction of the feasible solution set has a fundamental value. We propose a new methodology for systematically constructing the feasible solution set for a database search. This allows us to significantly improve the quality of the search results.

- Chapter 9 presents preliminary results of the application of the PSI method for the design of the L_1 flight control system implemented on the two-turbine-powered dynamically scaled Generic Transport Model (GTM), which is part of the Airborne Subscale Transport Aircraft Research aircraft at the NASA Langley Research Center [12]. In particular, the study addresses the construction of the feasible solution set and the improvement of a nominal prototype design, obtained using the systematic design procedures of the L_1 adaptive control theory. On one hand, the results of this chapter demonstrate the benefits of L_1 adaptive control as a verifiable robust adaptive control architecture by validating the theoretical claims in terms of robustness and performance, as well as illustrating its systematic design procedures. On the other hand, the developed procedure confirms the suitability of the PSI method for the multicriteria design optimization of a flight control system subject to multiple control specifications. Furthermore, in order to facilitate the multicriteria analysis process, this study takes advantage of the MOVI package, which was designed to apply the PSI method to engineering problems. The results and conclusions of this chapter have contributed to the improvement of the (predicted) flying qualities and the robustness margins of the all-adaptive L_1-augmented GTM AirSTAR aircraft. Chapter 9 is written in collaboration with Enrick Xargay and Vladimir Dobrokhodov.

Again, in Part II, our goal was to demonstrate multicriteria analysis of different real-life problems and to show how the expert controls constraints to find the optimal solution. All of these problems are united by the necessity of constructing the feasible solution set.

Examples of calculation of the approximate compromise curves are described in the appendix. The appendix considers two simple problems where compromise curves can be found by an analytical procedure; the calculated approximate compromise curves are compared to the exact ones.

In summary, the main idea of this book is to demonstrate that the construction and analysis of the feasible solution set are of primary importance in real-life optimization problems. The application of the PSI method and the MOVI software has allowed us to solve many problems that, until recently, appeared to be intractable. The expediency of further applications of various optimization methods, including stochastic, genetic, and other algorithms, depends first on the results of the analysis of feasible and Pareto optimal solutions obtained on the basis of the PSI method.

The application area of the PSI method is fairly large. In our book, we will be limiting ourselves to the construction and analysis of the feasible solution set as a basis of engineering optimization problems. We also address our book to all those who deal with real-life optimization in various areas of human activity, including biology, geology, chemistry, and physics [1–3, 5].

The authors of the MOVI project are V. K. Astashev, J. B. Matusov, M. N. Toporkov, K. S. Pyankov, A. R. Statnikov, R. B. Statnikov, and I. V. Yanushkevich. The project director is R. B. Statnikov.

This book provides an educational version of MOVI software on the enclosed disc. MOVI runs on Windows XP, Windows Vista, and Windows 7. The educational version can be used for education and noncommercial research only. Please check our Web site http://www.psi-movi.com for detailed installation instructions, technical support, and to download the latest version of MOVI. You can also contact the authors about commercial use of MOVI on our Web site.

References

[1] Statnikov, R. B., and J. B. Matusov, *Multicriteria Analysis in Engineering,* Dordrecht/Boston/London: Kluwer Academic Publishers, 2002.

[2] Statnikov, R. B., and J. B. Matusov, *Multicriteria Optimization and Engineering,* New York: Chapman & Hall, 1995.

[3] Statnikov, R. B., *Multicriteria Design. Optimization and Identification,* Dordrecht/Boston/London: Kluwer Academic Publishers, 1999.

[4] Statnikov, R. B., and J. B. Matusov, "Use of P_τ Nets for the Approximation of the Edgeworth-Pareto Set in Multicriteria Optimization," *Journal of Optimization Theory and Applications,* Vol. 91, No. 3, 1996, pp. 543–560.

[5] Sobol', I. M., and R. B. Statnikov, *Selecting Optimal Parameters in Multicriteria Problems,* 2nd ed., (in Russian), Moscow: Drofa, 2006.

[6] Lozino-Lozinsky, G. E., et al., "MAKS—Experimental Rocket Powered Plane Demonstrator of Technologies," *AIAA 7th Spaceplanes and Hypersonic Systems & Technology Conference,* Norfolk, VA, November 18–22, 1996.

[7] Lozino-Lozinsky, G. E., E. N. Dudar, and R. Joner, "Comparative Analysis of Reusable Space Transportation Systems," *31th AIAA/ASME/SAE/ASEE Joint Propulsion Conference,* San Diego, CA, 1995.

[8] Lozino-Lozinsky, G. E., V. A. Skorodelov, and V. P. Plokhikh, "International Reusable Aerospace System MAKS," *AIAA/DGLR 5th International Aerospace Planes and Hypersonic Technologies Conference,* Munich, Germany, November 30–December 3, 1993, pp. 18–22.

[9] Grumondz, V. T., and G. A. Yakovlev, *Algorithms for Aerohydroballistic Design,* (in Russian), Moscow: MAI, 1994.

[10] Yakimenko, O. A., and R. B. Statnikov, "On Multicriteria Parametric Identification of the Cargo Parafoil Model with the of PSI Method," P*roceedings of the 18th AIAA Aerodynamic Decelerator Systems Technology Conference and Seminary (AIAA 2005),* Munich, Germany, May 23–26, 2005.

[11] Dobrokhodov, V., and R. Statnikov, "Multi-Criteria Identification of a Controllable Descending System," *Proceedings of the First IEEE Symposium on Computational Intelligence in Multi-Criteria Decision-Making (MCDM 2007),* Honolulu, HI, 2007.

[12] Xargay, E., et al., "L_1 Adaptive Flight Control System: Systematic Design and V&V of Control Metrics," *AIAA Guidance, Navigation, and Control Cconference,* Toronto, Ontario, Canada, August 2–5, 2010.

[13] Egorov, I. N., et al., "Multicriteria Optimization of Complex Engineering Systems from the Design to Control," *Journal of Machinery Manufacture and Reliability,* Russian Academy of Sciences, No. 2, 1998, pp. 10–20.

[14] Gobbi, M., et al., "Multi-Objective Optional Design of Road Vehicle Sub-Systems by Means of Global Approximation," *Proceedings of the 15th European ADAMS Users' Conference,* 2000.

[15] Pavlov, Y. S., et al., "Multicriteria Simulation and Analysis," *Journal of Machinery Manufacture and Reliability,* Russian Academy of Sciences, No. 1, 1996, pp. 88–94.

[16] Bondarenko, M. I., et al., "Construction of Consistent Solutions in Multicriteria Problems of Optimization of Large Systems," *Physics-Doklady (Russian Academy of Sciences),* Vol. 39, No. 4, 1994, pp. 274–279.

[17] Chernykh, V. V., et al., "Parameter Space Investigation Method in Problems of Passenger Car Design," *Journal of Machinery Manufacture and Reliability, Russian Academy of Sciences,* Vol. 38, No. 4, 2009, pp. 329–334.

[18] Lur'e, Z. Y., A. I. Zhernyak, and V. P. Saenko, "Optimization of Pumping Units of Internal Involute Gear Pumps," *Journal of Machinery Manufacture and Reliability, Russian Academy of Sciences,* No. 3,1996, pp. 29–34.

[19] Statnikov, R. B., et al., "Visualization Approaches for the Prototype Improvement Problem," *Journal of Multi-Criteria Decision Analysis,* No. 15, 2008, pp. 45–61.

[20] Anil, K. A., "Multi-Criteria Analysis in Naval Ship Design," Master's Thesis, Naval Postgraduate School, Monterey, CA, 2005.

[21] Statnikov, R., et al., "Visualization Tools for Multicriteria Analysis of the Prototype Improvement Problem," *Proceedings of the First IEEE Symposium on Computational Intelligence in Multi-Criteria Decision-Making (MCDM 2007),* Honolulu, HI, 2007.

[22] Parsons, M. G., and R. L. Scott, "Formulation of Multicriterion Design Optimization Problems for Solution with Scalar Numerical Optimization Methods," *Journal of Ship Research,* Vol. 48, No. 1, 2004, pp. 61–76.

[23] Zverev, I. A., "Vector Identification of the Parameters of Spindle Units of Metal-Cutting Machines," *Journal of Machinery Manufacture and Reliability, Russian Academy of Sciences,* No. 6, 1997, pp. 58–63.

[24] Demyanushko, I. V., and M. E. Elmadawy, "Application of the Parameter Space Investigation Method for Optimization of Structures," *Transport Construction,* No. 7, 2009, pp. 26–28 (in Russian).

[25] Podgaets, A. R., and W. J. Ockels, "Problem of Pareto-Optimal Control for a High Altitude Energy System," *9th World Renewable Energy Congress WREC IX,* Florence, Italy, August 19–25, 2006.

[26] Podgaets, A. R., and W. J. Ockels, "Flight Control of the High Altitude Wind Power System," *Proceedings of the 7th Conference on Sustainable Applications for Tropical Island States,* Cape Canaveral, FL, June 3–6, 2007.

[27] Bordetsky, A., and R. Hayes-Roth, "Extending the OSI Model for Wireless Battlefield Networks: A Design Approach to the 8th Layer for Tactical Hyper-Nodes," *International Journal of Mobile Network Design and Innovation,* Vol. 2, No. 2, 2007, pp. 81–91.

[28] Bordetsky, A., et al., "Multicriteria Approach in Configuration of Energy Efficient Sensor Networks," *43rd Annual ACM Southeast Conference ACMSE,* Kennesaw, GA, Vol. 2, 2005, pp. 28–30.

[29] Ignat'ev, V. A., J. B. Matusov, and R. B. Statnikov, "Multicriteria Optimization of the Robot," *Journal of Machinery Manufacture and Reliability,* Russian Academy of Sciences, No. 5, 2000, pp. 75–83.

[30] Steuer, R. E., and M. Sun, "The Parameter Space Investigation Method of Multiple Objective Nonlinear Programming: A Computational Investigation," *Operations Research,* Vol. 43, No. 4, 2005, pp. 641–648.

[31] Lieberman, E., *Multi-Objective Programming in the USSR,* New York: Academic Press, 1992.

[32] Lieberman, E., "Soviet Multi-Objective Mathematical Programming Methods: An Overview," *Management Science,* Vol. 37, No. 9, 1991, pp. 1147–1165.

[33] Statnikov, R. B., et al., "Multicriteria Analysis Tools in Real-Life Problems," *Journal of Computers & Mathematics with Applications,* Vol. 52, 2006, pp. 1–32.

[34] Statnikov, R. B., A. Bordetsky, and A. Statnikov, "Multicriteria Analysis of Real-Life Engineering Optimization Problems: Statement and Solution," *Nonlinear Analysis,* Vol. 63, 2005, pp. e685–e696.

[35] Statnikov, R., et al., "Definition of the Feasible Solution Set in Multicriteria Optimization Problems with Continuous, Discrete, and Mixed Design Variables," *Nonlinear Analysis,* Vol. 71, No. 12, 2009, pp. e109–e117.

[36] Statnikov, R., A. Bordetsky, and A. Statnikov, "Management of Constraints in Optimization Problems," *Nonlinear Analysis,* Vol. 71, No. 12, 2009, pp. e967–e971.

[37] Statnikov, R. B., et al., *MOVI 1.4 (Multicriteria Optimization and Vector Identification) Software Package,* Certificate of Registration, Register of Copyrights, U.S.A., Registration Number TXU 1-698-418, Date of Registration: May 22, 2010.

[38] Wasserman, P. D., *Advanced Methods in Neural Computing,* New York: Van Nostrand Reinhold, 1993.

[39] Collobert, R., and S. Bengio, "SVMTorch: Support Vector Machines for Large-Scale Regression Problems," *Journal of Machine Learning Research,* February 1, 2001, pp. 143–160.

[40] Anderson, T. W., *An Introduction to Multivariate Statistical Analysis,* 3rd ed., New York: Wiley-Interscience, 2003.

[41] Chatterjee, S., and A. S. Hadi, "Influential Observations, High Leverage Points, and Outliers in Linear Regression," *Statistical Science,* Vol. 1, No. 3, 1986, pp. 379–416.

[42] Vapnik, V., *Statistical Learning Theory,* New York: John Wiley & Sons, 1998.

[43] Statnikov, R. B., et al., "DBS-PSI: A New Paradigm of Database Search," *International Journal of Services Sciences,* Vol 4, No. 1, 2011, pp. 1–13.

Part I
The Parameter Space Investigation Method Toolkit

For the correct formulation and solution of real-life optimization problems, a method called the parameter space investigation (PSI) has been created and widely integrated into various fields of human activity. The PSI method is implemented in the Multicriteria Optimization and Vector Identification (MOVI) software system.

1

Introduction

1.1 Some Basic Features of Real-Life Optimization Problems

Let us start by enumerating some basic features of real-life optimization problems.

- These problems are essentially multicriteria ones. Numerous attempts to reduce multicriteria problems to single-criterion problems have proved to be fruitless. By presenting a multicriteria problem in the form of a single-criterion problem, we replace the initial problem with a different problem that has little in common with the original one. Obviously, one should always try to take into account all basic performance criteria simultaneously.
- The determination of the feasible solution set is one of the fundamental issues of the real-life problems. The construction of the feasible set is an importance step in the formulation and solution of such problems.
- As a rule, the feasible solution set may be obtained only in the process of solving a problem and analysis of results. Analysis of the feasible set allows one to not only correct the initial constraints, but also to revise the original mathematical models and list of criteria. The multicriteria problems should be formulated and solved in the interactive mode. The formulation and solution of a real-life problem should be a single process [1].

- Mathematical models can be complex systems of equations (including differential and other types of equations) that may be linear or nonlinear, deterministic, or stochastic, with distributed or lumped parameters. Sometimes mathematical models have to be derived from observational data using statistical machine learning techniques.
- The feasible solution set can be multiply connected, and its volume may be several orders of magnitude smaller than that of the domain within which the optimal solution is sought.
- Both the feasible solution set and the Pareto optimal set are nonconvex in the general case. As a rule, information about the smoothness of objective functions and functional dependences is absent. These functions are usually nonlinear; they may be nondifferentiable as well.
- A real-life optimization problem may contain a large number of constraints, and the dimensionality can reach: (1) many hundreds and thousands for the design variable vector, and (2) many dozens, if not hundreds, for the criterion vector.
- Very often experts do not encounter serious difficulties in analyzing the Pareto optimal set and in choosing the most preferred solution [1–5]. This is because experts have a sufficiently well-defined system of preferences[1] in this type of problems and the Pareto optimal set often contains a small number of solutions due to stringent constraints.

To formulate and solve real-life optimization problems, a method called the parameter space investigation (the PSI method) has been developed. A systematic and comprehensive description of this method can be found in [1, 5].

1.2 Generalized Formulation of Multicriteria Optimization Problems

We assume that an object depends on r design variables $\alpha_1, \ldots, \alpha_r$ representing a point $\alpha = (\alpha_1, \ldots, \alpha_r)$ in the r-dimensional space. Generally, an expert has to take into account the design variable, functional, and criteria constraints.

The *design variable constraints* have the form $\alpha_j^* \le \alpha_j \le \alpha_j^{**}$, $j = 1, \ldots, r$. In the case of mechanical systems, the α_j represent stiffness coefficients, the moments of inertia, damping factors, geometric sizes, and so forth.

The *functional constraints* can be written as follows: $C_l^* \le f_l(\alpha) \le C_l^{**}$, $l = 1, \ldots, t$, where $f_l(\alpha)$ is a functional dependence (relation), C_l^* and C_l^{**} are

1. More complex cases of the decision making, where preferences on the Pareto optimal set are not necessarily stable, are discussed in [6].

some constants. Functional constraints can be the standards, allowable stress in structural elements, and other requirements to the system.

The design variable constraints single out a parallelepiped Π in the r-dimensional design variable space (space of design variables); see Figure 1.1. In turn, design variable and functional constraints together define a certain subset G in Π; see Figure 1.2.

There also exist particular performance criteria, such as fuel consumption, efficiency, cost, and so on. It is desired that, with other things being equal, these criteria, denoted by $\Phi_v(\alpha)$, $v = 1, \ldots, k$, would reach extremal values. We assume that $\Phi_v(\alpha)$ are to be minimized.

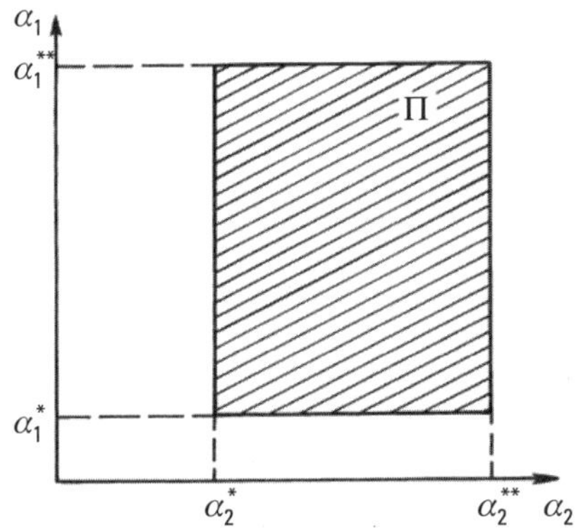

Figure 1.1 Parallelepiped Π.

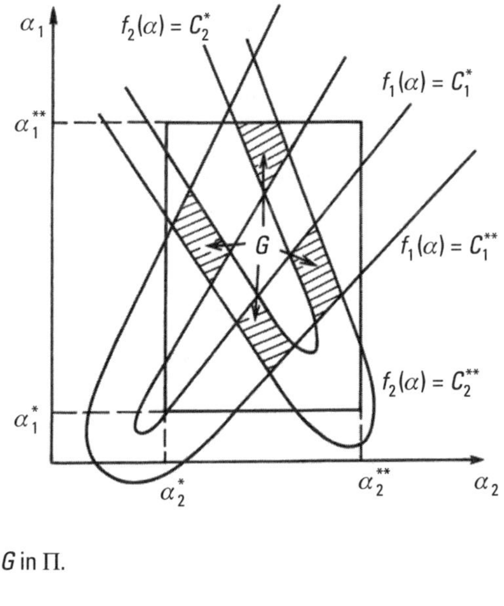

Figure 1.2 Subset G in Π.

Unlike criteria, functional dependences do not need to be optimized.

To avoid situations in which the expert regards the values of some criteria as unacceptable, we introduce *criteria constraints* in the form $\Phi_v(\alpha) \le \Phi_v^{**}$, $v = 1, \ldots, k$, where Φ_v^{**} is the worst value of criterion $\Phi_v(\alpha)$ acceptable to an expert. The choice of Φ_v^{**} is discussed in the Chapter 2.

The design variable, functional, and criteria constraints define the *feasible solution set* $D \subset G \subset \Pi$; see Figure 1.3.

If functions $f_l(\alpha)$ and $\Phi_v(\alpha)$ are continuous in Π, then the sets G and D are closed.

Let us formulate one of the basic problems of multicriteria optimization. It is necessary to define the feasible solution set D and find a set $P \subset D$ such that

$$\Phi(P) = \min_{\alpha \in D} \Phi(\alpha) \tag{1.1}$$

where $\Phi(\alpha) = (\Phi_1(\alpha), \ldots, \Phi_k(\alpha))$ is the criterion vector and P is the *Pareto optimal set.* We mean that $\Phi(\alpha) < \Phi(\beta)$ if for all $v = 1, \ldots, k$, $\Phi_v(\alpha) \le \Phi_v(\beta)$ and for at least one $v_0 \in \{1, \ldots, k\}$, $\Phi_{v_0}(\alpha) < \Phi_{v_0}(\beta)$.

When solving this problem, one has to determine the vector of design variables $\alpha^O \in P$, which is the most preferable one among the vectors belonging to set P.

Let us give an alternative definition of the Pareto optimal set.

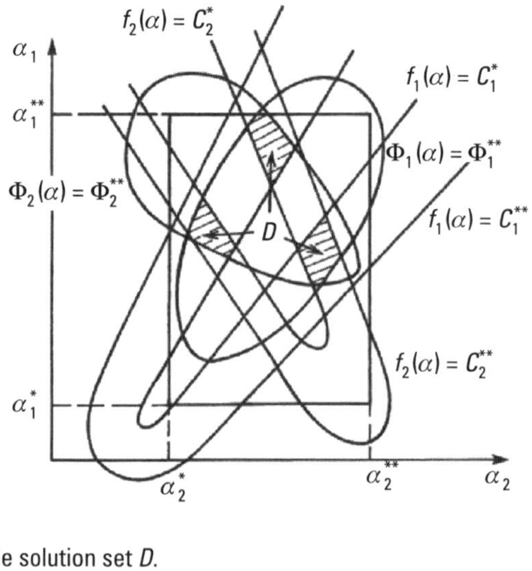

Figure 1.3 Feasible solution set *D*.

1.2.1 Definition

A point $\alpha^O \in D$ is called the Pareto optimal point if there exists no point $\alpha \in D$ such that $\Phi_v(\alpha) \leq \Phi_v(\alpha^O)$ for all $v = 1,\ldots k$ and $\Phi_{v_0}(\alpha) < \Phi_{v_0}(\alpha^O)$ for at least one $v_0 \in \{1, \ldots, k\}$. A set $P \subset D$ is called a Pareto optimal set if it consists of Pareto optimal points.

The Pareto optimal set plays an important role in multicriteria optimization problems because it can be analyzed more easily than the feasible solution set and because the most preferable solution always belongs to the Pareto optimal set, irrespective of the system of preferences used by the expert for comparing vectors belonging to the feasible solution set [1–5, 7–10].

1.3 Applying Single-Criterion Methods for Solving Multicriteria Problems

As already noted, the real-life problems are essentially multicriteria. However, in an overwhelming majority of cases, these problems are solved as single ones. Consider two widespread approaches.

1.3.1 Substitution of a Multitude of Criteria by a Single One

For instance, in this case one needs to choose weight coefficients $\beta_v \geq 0$ (usually, $\beta_1 + \ldots + \beta_k = 1$) so that the function

$$\Phi(\alpha) = \beta_1\Phi_1(\alpha) + \ldots + \beta_k\Phi_k(\alpha)$$

integrates all criteria $\Phi_1,\ldots, \Phi_k$ and to consider $\Phi(\alpha)$ as the only performance criterion. The coefficient β_v reflects the relative "importance" of the criterion Φ_v, $v = 1,\ldots, k$.

In practice, the values of β_v are usually unknown beforehand, especially if these criteria are of different natures and reflect different aspects of the system's behavior.

To use this approach we should answer the most challenging questions:

- How do we define a decision rule (or how do we convolute the criteria)? In other words, why should we sum the criteria instead of multiplying them or carrying out other operations?
- How do we define the weight coefficients?

Also, before convoluting the criteria, we should normalize them (i.e., make dimensionless). Notice that normalization of criteria for solving a specific problem is a difficult task.

1.3.2 Optimization of the Most Important Criterion

In this case the criterion $\Phi_1(\alpha)$ considered by the expert to be the most important is optimized, while all the others are replaced by constraints. We have to choose constraints $\Phi_2^{**},\dots,\Phi_k^{**}$ and consider the problem of finding the minimum

$$\Phi_1(\alpha) \rightarrow \min$$

under the following constraints:

$$\begin{aligned} &\alpha_j^* \le \alpha_j \le \alpha_j^{**}, && j = 1,\dots,r, \\ &C_l^* \le f_l(\alpha) \le C_l^{**}, && l = 1,\dots,t, \\ &\Phi_v(\alpha) \le \Phi_v^{**}, && v = 2,\dots,k. \end{aligned}$$

In this approach we should answer the challenging question: What is the most important criterion? Notice that the majority of real-life problems contain several meaningful criteria, and some of them are conflicting. This is a characteristic feature of real-life problems.

In general, in cases 1.3.1 and 1.3.2 it is difficult, sometimes even impossible to answer the above-mentioned questions correctly. However, if we even answer these questions, the main question, namely, how to construct the feasible solution set D (to define constraints $\alpha_j^*, \alpha_j^{**}, C_l^*, C_l^{**}, \Phi_v^{**}$), still remains open, that is, there is no guarantee that we will search for the optimal solutions in the right place. With it we do not exclude an opportunity of representation of multicriteria problems as single-criterion ones. However, preliminarily it is necessary to define the feasible solution set. The problem of correct constructing the feasible solution set will be discussed in the Chapters 2 and 3.

We will revisit and cover in greater detail the substitution of a multitude of criteria by a single one in Sections 2.2 and 2.3.

1.4 Systematic Search in Multidimensional Domains by Using Uniformly Distributed Sequences

The features of the problems under consideration make it necessary to represent vectors α by points of uniformly distributed sequences in the design variable space [2, 4, 5]. We briefly summarize this approach next.

For many applied problems the following situation is typical. There exists a multidimensional domain in which a function (or a system of functions) is considered whose values may be calculated at certain points. Suppose we wish to get some information on the behavior of the function in the entire domain or in any subdomain. Then, in the absence of any additional information about the function, it is natural to wish that the points at which the function is calculated would be uniformly distributed within the domain. However, the following question arises: What meaning should be assigned to the notion of a uniform distribution? This concept is quite evident only in the case of a single variable. By dividing the range of the variable into N equal parts and locating a point within each of the parts, we arrive at a sequence of N points (a net) uniformly distributed over the domain under consideration. Unfortunately, in the case of several variables the concept of uniformity is not so evident. If for each of the variables we make a partition similar to that done in the case of a single variable, then for n variables we get N^n points (a cubic net). However, the concept of uniformity should be independent of the number of points, and, in addition, the use of nets containing so many points seriously complicates the solution of practical problems.

Weyl was the first to give the definition of uniformity [5, 7]. Let us consider a sequence of points $P_1, P_2, \ldots, P_i, \ldots$ belonging to a unit r-dimensional cube K^r. We denote by G an arbitrary domain in K^r and we denote by $S_N(G)$ the number of points P_i belonging to G $(l \leq i \leq N)$. The sequence P_i is called *uniformly distributed* in K^r, if

$$\lim_{N \to \infty} \frac{S_N(G)}{N} = V_G \tag{1.2}$$

where V_G is the volume of the r-dimensional domain G.

The meaning of the definition is the following [1, 5, 7]: for large values of N, the number of points of a given sequence belonging to an arbitrary domain G is proportional to volume V_G:

$$S_N(G) \sim NV_G \tag{1.3}$$

In solving real-life problems, one must commonly deal not with K^r, but with a certain parallelepiped Π, and, hence, move from the coordinates of the points uniformly distributed in K^r to those in Π.

Let us formulate the following statements [5]. If points Q_i with Cartesian coordinates $(q_{i1}, ..., q_{ir})$ form a uniformly distributed sequence in K^r, then points α^i with Cartesian coordinates $\alpha_1^i,\dots,\alpha_r^i$, $1 \le i \le N$, where

$$\alpha_j^i = \alpha_j^* + q_{ij}\left(\alpha_j^{**} - \alpha_j^*\right),\ j = 1,2,\dots,r,\ 0 < q_{ij} < 1 \tag{1.4}$$

form a uniformly distributed sequence in parallelepiped Π consisting of points $(\alpha_1^i,\dots,\alpha_r^i)$ whose coordinates satisfy the inequalities $\alpha_j^* \le \alpha_j \le \alpha_j^{**}$.

Let $\alpha^1,..., \alpha^i, ...$ be a sequence of points uniformly distributed in Π, and $G \subset \Pi$ be an arbitrary domain with volume $V_G > 0$. If among the points α^i, one chooses all the points belonging to G, then one obtains the sequence of points uniformly distributed in G [5].

1.4.1 Quantitative Characteristics of Uniformity

Let us fix a net consisting of the points $P_1, ..., P_N \in K$. To estimate the uniformity of distribution of these points quantitatively, we introduce the quantity $D(P_1, ..., P_N)$, called the *discrepancy*, implying the discrepancy between the "ideal" and actual uniformities.

Let P be an arbitrary point belonging to K and G_P be an n-dimensional parallelepiped with the diagonal OP and faces parallel to the coordinate planes (Figure 1.4). Denote by V_{G_P} the volume of G_P and by $S_N(G_P)$, the number of points P_i that enter G_P and whose subscripts satisfy the inequalities $1 \le i \le N$.

The discrepancy of the points $P_1, ..., P_N$ is the number

$$D(P_1,\dots,P_N) = \sup_{P \in K}\left|S_N(G_P) - NV_{G_P}\right| \tag{1.5}$$

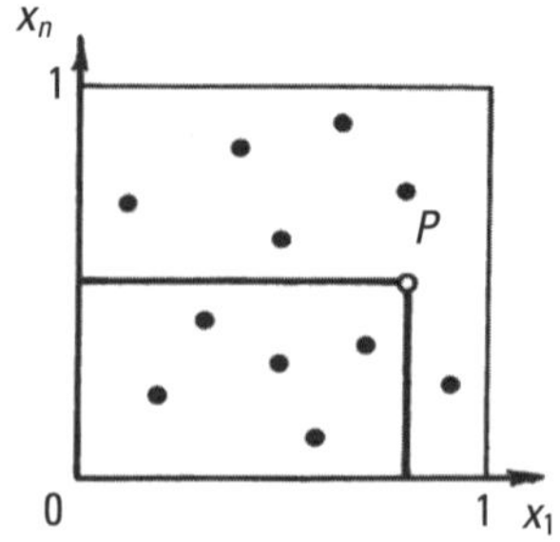

Figure 1.4 Determination of the discrepancy.

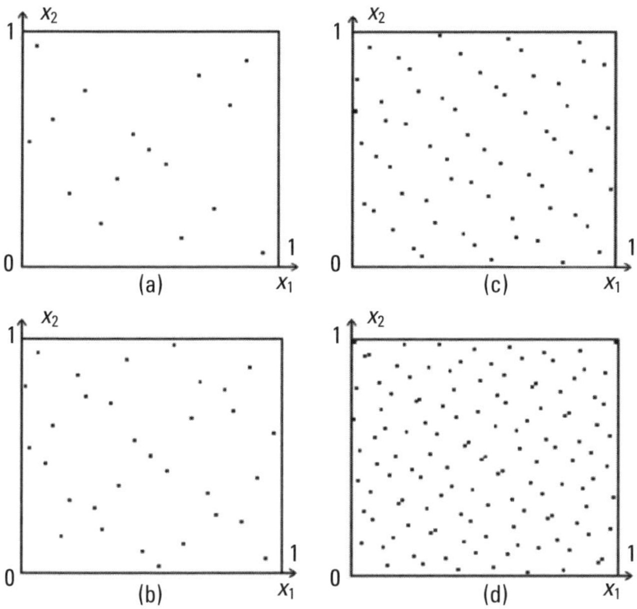

Figure 1.5 Points Q_i in K^2: (a), (b), (c), and (d) correspond to $N = 16, 32, 64$, and 128.

where the supremum is taken over all possible positions of the point P in the cube. It is natural to consider that the smaller $D(P_1, ..., P_N)$ is, the more uniformly the points $P_1, ..., P_N$ are arranged. We mention here works by Halton [11], Hammersley [12], Hlawka [13], Faure [14], and Kuipers and Niederreiter [15, 16], in which uniformly distributed sequences and nets (in the sense of the uniformity estimates) have been constructed.

Among uniformly distributed sequences known at present, the so-called LP_τ sequences are among the best ones as regards uniformity characteristics as $N \to \infty$, see [1–5]. The points of LP_τ sequences $Q_i = (q_{i1}; q_{i2})$, $i = \overline{1, N}$ are shown in Figure 1.5.

References

[1] Statnikov, R. B., and J. B. Matusov, *Multicriteria Analysis in Engineering,* Dordrecht/Boston/London: Kluwer Academic Publishers, 2002.

[2] Statnikov, R. B., and J. B. Matusov, *Multicriteria Optimization and Engineering,* New York: Chapman & Hall, 1995.

[3] Statnikov, R. B., *Multicriteria Design: Optimization and Identification,* Dordrecht/Boston/London: Kluwer Academic Publishers, 1999.

[4] Statnikov, R. B, and J. B. Matusov, "Use of P_τ Nets for the Approximation of the Edgeworth-Pareto Set in Multicriteria Optimization," *Journal of Optimization Theory and Applications*, Vol. 91, No. 3, 1996, pp. 543–560.

[5] Sobol', I. M., and R. B. Statnikov, *Selecting Optimal Parameters in Multicriteria Problems*, 2nd ed., (in Russian), Moscow: Drofa, 2006.

[6] Lichtenstein, S., and P. Slovic, *The Construction of Preference*, New York: Cambridge University Press, 2006.

[7] Weyl, H., "Uber die Gleichverteilung von Zahlen mod. Eins," *Math. Ann.*, Vol. 77, 1916, pp. 313–352.

[8] Statnikov, R. B., et al., "Multicriteria Analysis Tools in Real-Life Problems," *Journal of Computers and Mathematics with Applications*, Vol. 52, 2006, pp. 1–32.

[9] Statnikov, R. B., A. Bordetsky, and A. Statnikov, "Multicriteria Analysis of Real-Life Engineering Optimization Problems: Statement and Solution," *Nonlinear Analysis*, Vol. 63, 2005, pp. e685–e696.

[10] Statnikov, R., A. Bordetsky, and A. Statnikov, "Management of Constraints in Optimization Problems," *Nonlinear Analysis*, Vol. 71, No. 12, 2009, pp. e967–e971.

[11] Halton, J. H., "On the Efficiency of Certain Quasi-Random Sequences of Points in Evaluating Multi-Dimensional Integrals," *Numerische Mathematik*, Vol. 2, 1960, pp. 84–90.

[12] Hammersley, J. M., "Monte Carlo Methods for Solving Multivariable Problems," *Ann. New York Acad. Sci.*, No. 86, 1960, pp. 844–874.

[13] Hlawka, E., and R. Taschner, *Geometric and Analytic Number Theory*, Berlin, Germany: Springer, 1991.

[14] Faure, H., "Discrepancy of Sequences Associated with a Number System (in Dimensions)," *Acta Arithmetica*, Vol. 41, 1982, pp. 337–351, in French.

[15] Kuipers, L., and H. Niederreiter, *Uniform Distribution of Sequences*, New York: John Wiley & Sons, 1974.

[16] Niederreiter, H., "Statistical Independence Properties of Pseudorandom Vectors Produced by Matrix Generators," *J. Comput. and Appl. Math.*, No. 31, 1990, pp. 139–151.

2

Parameter Space Investigation Method as a Tool for Formulation and Solution of Real-Life Problems

The PSI method fully meets most important features of engineering optimization problems, including problems with "soft" functional constraints.

We will consider why correct statement of optimization problem is very important and why an expert requires help for definition of the feasible solution set (constraints on design variables, functional dependences, and criteria).

The greater the number of criteria taken into account, the greater the information obtained about: (1) the resources of improving the object (ship, car, nuclear reactor, aircraft, machine tool, robot, submarine); (2) the performance of a mathematical model and constraints; and (3) the accuracy with which the criteria are calculated and how much one can trust them.

The PSI method allows us to consider as many criteria as necessary.

The material described in this chapter is provided in full detail in many references [1–9]. That is why we confined ourselves to the brief description of the basis of the PSI method.

2.1 The Parameter Space Investigation Method

In Section 1.2 we formulated the problem of multicriteria optimization and defined the feasible solution set D, which is constructed using the values of $\mathbf{\Phi}_v^{**}$, $v = 1,\ldots, k$ and other constraints. Now we proceed by describing the parameter space investigation (PSI) method that allows us to determine $\mathbf{\Phi}_v^{**}$ and, hence, the feasible solution set.

The method consists of three stages (see Figure 2.1).

Stage 1: Compilation of Test Tables Via Computer

First, N trial points $\alpha^1, \ldots, \alpha^N$ that satisfy the functional constraints are generated. Then all the particular criteria $\Phi_v(\alpha^i)$ are calculated at each of the points α^i; for each of the criteria a test table is compiled so that the values of $\Phi_v(\alpha^1), \ldots, \Phi_v(\alpha^N)$ are arranged in increasing order; that is,

$$\Phi_v\left(\alpha^{i_1}\right) \le \Phi_v\left(\alpha^{i_2}\right) \le \ldots \le \Phi_v\left(\alpha^{i_N}\right), v = 1, \ldots, k \tag{2.1}$$

where $i_1, i_2, \ldots, i_N$ are the numbers of trials (a separate set for each v). Taken together, the k tables form complete test tables. In other words, all values of each criterion are arranged in a test table in order (from best to worst values).

Stage 2: Selection of Criterion Constraints

This stage includes dialogs of an expert with computer. By analyzing inequalities (2.1), an expert specifies the criteria constraints Φ_v^{**}. Actually, an expert has to consider one criterion at a time and specify the respective constraints. He analyzes one test table and imposes the criterion constraint. Then he proceeds to the next test table, and so on. Note that the revision of the criteria constraints does not lead to any difficulties for an expert.

Since we want to minimize all criteria, Φ_v^{**} should be the maximum values of the criteria $\Phi_v(\alpha)$ which guarantee an acceptable level of the object's operation.

If the selected values of Φ_v^{**} are not a maximum, then many interesting solutions may be lost, since some of the criteria are contradictory. Moreover, in some cases, the feasible solution set may be empty.

In practice the expert imposes the criteria constraints in order to improve a prototype by all criteria simultaneously. If it is impossible, he or she improves a prototype by the most important criteria. In process of dialogues with computer, the expert repeatedly revises criteria constraints and carries out the multicriteria analysis. The PSI method gives the expert valuable information on the advisability of revising various criteria constraints with the aim of improving the basic criteria. The expert sees what price one pays for making concessions in various criteria (i.e., what one loses and what one gains).

Stage 3: Verification of the Solvability of Problem Via Computer

Let us fix a criterion, for example, $\Phi_{v_1}(\alpha)$, and consider the corresponding test table (2.1). Let S_1 be the number of the values in the table satisfying the selected criterion constraint:

$$\Phi_{v_1}\left(\alpha^{i_1}\right) \le \ldots \le \Phi_{v_1}\left(\alpha^{i_{S_1}}\right) \le \Phi_{v_1}^{**} \tag{2.2}$$

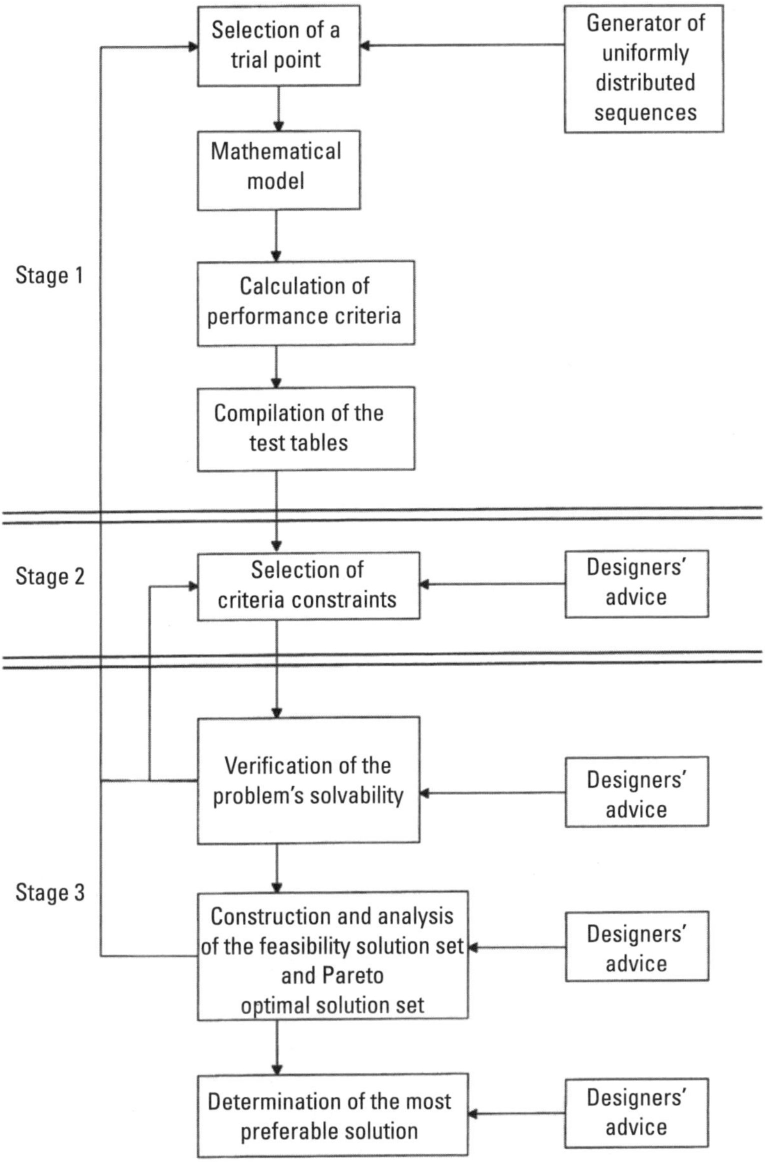

Figure 2.1 Flowchart of the algorithm.

Then criterion Φ_{v_2} is selected by analogy with Φ_{v_1} and the values of $\Phi_{v_2}(\alpha^{i_1}),\ldots,\Phi_{v_2}(\alpha^{i_{S_1}})$ in the test table are considered. Let the table contain $S_2 \le$

S_1 values such that $\Phi_{v_2}(\alpha^{i_j}) \leq \Phi_{v_2}^{**}$ where $1 \leq j \leq S_2$. Similar procedures are carried out for each criterion. Then if at least one point can be found for which all criteria constraints are valid simultaneously, then the set D is nonempty and problem (1.1) is solvable. Otherwise, the expert should return to Stage 2 and make certain concessions in criteria constraints Φ_v^{**}. However, if the concessions are highly undesirable, then one may return to Stage 1 and increase the number of points N in order to repeat Stage 2 and Stage 3.

The procedure is iterated until D is nonempty. Then the Pareto optimal set is constructed in accordance with the definition presented above. This is done by removing those feasible points that can be improved with respect to all criteria simultaneously.

Thus, in accordance with the PSI method, the criteria constraints are determined in the dialogue of an expert with a computer. Then an expert should determine the Pareto set P and after analyzing P, find the most preferred solution $\Phi(\alpha^0)$. As already noted, for discussed problems experts do not encounter serious difficulties in analyzing the Pareto optimal set and in choosing the most preferred solution.

The approximation of the feasible solution and Pareto optimal sets with a given accuracy is considered in [1, 2, 4]. In the appendix examples of calculation of the approximate compromise curves are shown. The definition of the feasible solution set in multicriteria problems with discrete and mixed design variables on the basis of PSI method is described in [6].

2.1.1 The Complexity of the Investigation

The property of uniform distribution of points implies that $\gamma = V(D)/V(\Pi) \approx N_D/N$ for sufficiently large N. Here N is the number of points $\alpha^i \in \Pi$; N_D is the number of points that have entered the feasible solution set; $V(D)$ is a volume of the feasible solution set; $V(\Pi)$ is a volume of the parallelepiped Π. The ratio of the volumes γ in the Monte Carlo theory is called the selection efficiency. For many engineering problems $\gamma << 0.001$; thus the search for feasible solution is like looking for a needle in a haystack. In fact, γ characterizes the complexity of solving problems. Very often after correction of constraints, the value of γ is vastly increased. The number of tests in the real-life problems strongly depends on a correctness of the initial statement of a problem. Using the PSI method, most of the computer time is spent on determining the feasible solution set, the correction of the problem statement that eventually leads to obtaining the justified optimal solutions. The number of trials in the real-life problems will be considered in Chapter 7.

The PSI method has proved to be a very convenient and effective tool for the expert, primarily because this method can be used for the statement and solution of the problem in an interactive mode.

2.1.2 Definition of the Feasible Solution in Parallel Mode

In some cases the time necessary for construction of the feasible solution set is one of the most important factors (e.g., in investigating high-dimensional problems). The use of the MOVI allows us to carry out optimization on many computers simultaneously, which obviously takes far less time than if we had tried to solve these problems on a single computer. In this case, the number of tests N can be distributed between k computers. So each computer finds a feasible solution set for its own subproblem. Then it is necessary to combine feasible solutions and construct a combined Pareto optimal set (see Section 4.2 and Chapter 7).

2.1.3 Number Generators for Systematic Search in the Design Variable Space

To investigate design variable space, we use *uniformly distributed sequences.* We often prefer the LP_τ sequences. The points of LP_τ sequences $Q_i = (q_{i,1}, q_{i,2}, ..., q_{i,50})$, belonging to a unit cube K^{50}, $\boldsymbol{i} = \overline{\mathbf{1}, \boldsymbol{N}}$ for $N = 256$, 2,048, 4,096, and 8,192 are shown in Figure 2.2. (These projections were constructed using MOVI.)

LP_τ sequences are used to compute N test points $\alpha^1,..., \alpha^N$ in the design variable space during Stage 1 of the PSI method. Other uniformly distributed sequences and nets [4, 10–15] can be also successfully used in the PSI method, see scheme presented in Figure 2.1.

However, prior to solving a specific problem, one cannot say with certainty which uniformly distributed sequences or nets are the most suitable. Much depends on the behavior of the criteria, the form of the functional and design variable constraints, the number of tests, and the geometry of the feasible solution set. Steuer and Sun indicated an opportunity of using random number generators in the PSI method [16]. Based on these recommendations, we have successfully applied random number generators in the PSI method to solve multicriteria problems with very high-dimensional design variable vectors (dimensionality = 1,000) and to also solve the problems in the parallel mode [7, 8]. The experiments with various random number generators in the PSI method are described in [9].

2.2 "Soft" Functional Constraints and Pseudo-Criteria

Using the traditional approach to multicriteria problems, one tries to reduce the number of criteria. The criticism of such approach is described in Section 1.2.

From the standpoint of our technique, it is necessary to act on the contrary. As indicated in the preface, we have two kinds of functional constraints: rigid and "soft." Usually "soft" functional constraints can be changed, if these

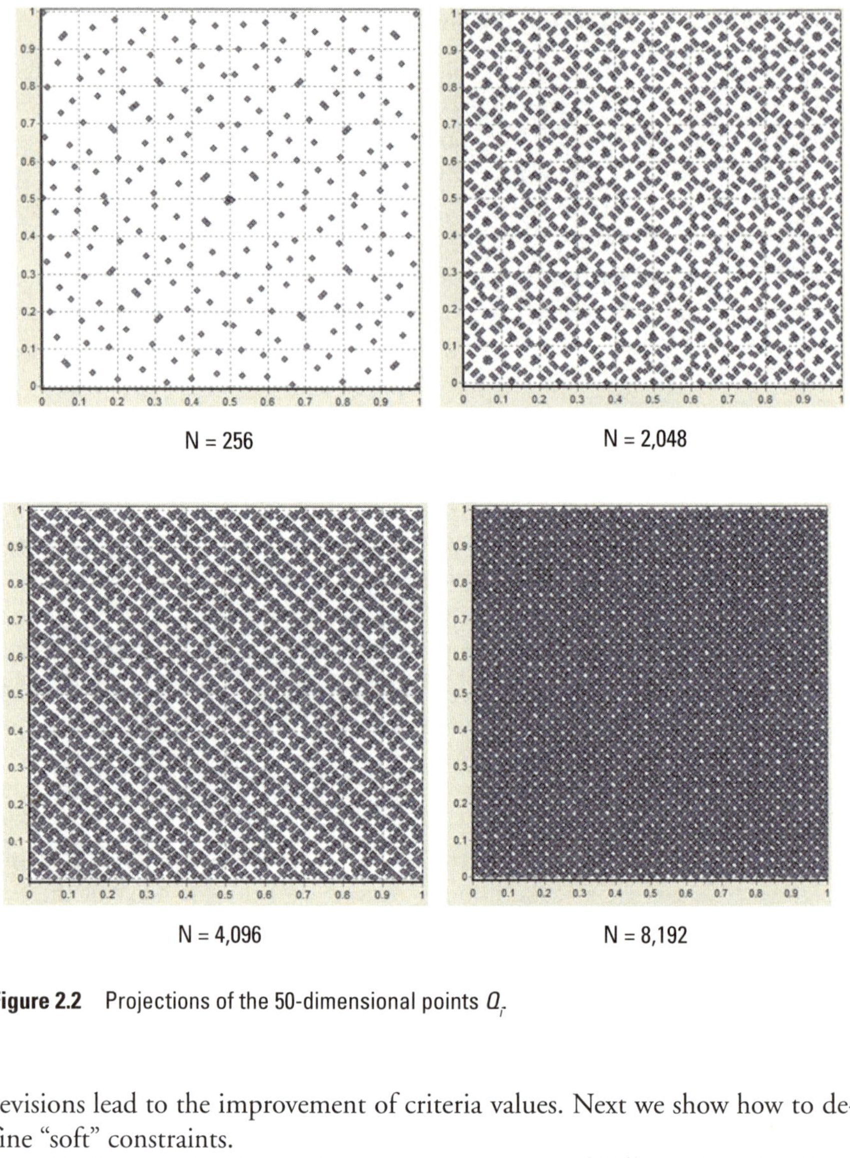

Figure 2.2 Projections of the 50-dimensional points Q_i.

revisions lead to the improvement of criteria values. Next we show how to define "soft" constraints.

In the case of unjustifiably strong constraints $C_l^*(C_l^{**})$ on functional dependences, many solutions can become unfeasible (i.e., do not satisfy criteria constraints Φ_v^{**}). For this reason the feasible solution set can be poor or even empty. Therefore, it is very important to help the expert determine the "soft" functional constraints correctly.

Let us assume that $f_l(\alpha) \le C_l^{**}$, $l = 1, \ldots, t$, where C_l^{**} are the "soft" constraints. The concept of pseudo-criteria is presented as the following. Instead of the function $f_l(\alpha)$, we introduce an additional criterion $\Phi_{k+l}(\alpha) = f_l(\alpha)$, which we call a pseudo-criterion. To find the value of the constraint Φ_{k+l}^{**}, one has to compile a test table containing $\Phi_{k+l}(\alpha)$. By using the PSI method, one can define Φ^{**}_{k+l} in a way that prevents the loss of interesting solutions. (Obviously, if rigid constraints under certain conditions can be changed, then the corresponding functional dependences should be also presented as pseudo-criteria.) In general, when solving a problem with "soft" functional constraints, one has to find the set D, taking all performance criteria and pseudo-criteria into account. In other words, one must solve the problem with the constraints

$$\Phi_v(\alpha) \leq \Phi_v^{**}, v = 1,\ldots,k,\ k+1,\ldots,\ k+t$$

Thus, to define the feasible solution set, we consider a multicriteria problem with $k + t$ criteria (and pseudo-criteria). Notice, however, that the pseudo-criteria are not considered when constructing the Pareto optimal set.

It is worthwhile to mention that many single-criterion problems have "soft" functional constraints (e.g., weight, dimensions, longevity, and so on). In these cases the definition of the feasible solution set is also very important. For determining this set, it is necessary to represent functional dependences as pseudo-criteria. In other words, we have to consider such single-criterion problems as multicriteria ones. In these cases, the test tables contain one performance criterion and the rest are pseudo-criteria.

2.3 More About Applying Single-Criterion Methods for Solving Multicriteria Problems

When an expert optimizes a single criterion from the set of criteria, he or she frequently does not know what the values of other important criteria are. Quite often, they turn out to be incorrect because of the imperfection of the mathematical models. Using single-criterion methods, the expert does not have sufficient information about the feasible solution set.

Often constraints that should define the feasible solution set in single-criteria problems are not feasible. As a result, there is no guarantee that the found optimal solution is feasible. In addition to this, single-criterion methods are not intended to search for compromise solutions or to determine the preferable solution among them.

The more criteria are optimized, the more information the expert obtains about compromise solutions, the model's work, and the correctness of calculating criteria. The expert obtains this information using the PSI method.

As noted earlier, in the case of optimizing a convolution of criteria, the expert is facing intractable problems of: (1) a substantiation of convolution, (2) normalization of criteria, and (3) definition of weight factors. However, we do not have to deal with these problems when using the PSI method. In our case the expert uses nonnormalized values of criteria (in pounds, dollars, inches, gallons, hours, mph, and so on) in the test table, so it is easy to analyze the obtained results and make decisions.

In Section 1.3 we considered the two typical statements of single-criterion problem. As indicated earlier, in these cases the complex problem of determining the feasible solution set is shifted onto the expert's shoulders. Unfortunately, the expert is usually unable to do this. The PSI method gives the opportunity to define the feasible solution set. Only after constructing and analyzing the feasible solution set and Pareto optimal set the expert can consider applying single-criterion optimization methods for the further improvement of the obtained results (e.g., the value of any criterion).

2.4 An Example of Optimization Problem Statement and Significant Challenge That It Presents

Here we consider optimization problem statement and the typical difficulties encountered at this stage.

The vibratory system consists of two bodies with masses M_1 and K_2 (see Figure 2.3).

- The mass M_1 is attached to a fixed base by a spring with stiffness coefficient K_1.
- A spring-and-dashpot[1] element with stiffness coefficient K_2 and damping coefficient C is located between masses M_1 and M_2.
- The harmonic force acts upon mass M_1.
- The amplitude and frequency of the exciting force are identified as $P =$ 2,000 (N) and $\omega = 30$ (s^{-1}).

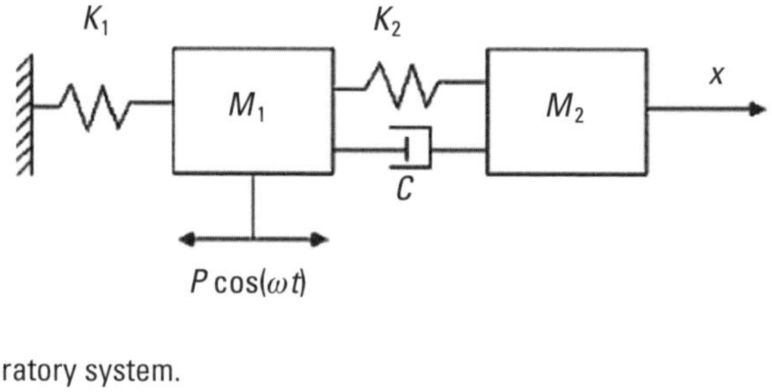

Figure 2.3 Vibratory system.

1. A dashpot is a mechanical device, a damper that resists motion via friction. The resulting force is proportional to the velocity, but acts in the opposite direction, slowing the motion and absorbing energy. It is commonly used in conjunction with a spring (which acts to resist displacement).

The motion of this system is governed by the equations:

$$M_1X_1''+C(X_1'-X_2')+K_1X_1+K_2(X_1-X_2)=P\cdot\cos(\omega t)$$
$$M_2X_2''+C(X_2'-X_1')+K_2(X_2-X_1)=0 \qquad (2.3)$$

We treat the parameters K_1, K_2, M_1, M_2, and C as the design variables to be determined, that is:

$$\alpha_1 = K_1, \alpha_2 = K_2, \alpha_3 = M_1, \alpha_4 = M_2, \alpha_5 = C$$

The design variable constraints are prescribed as the parallelepiped Π defined by the inequalities:

$$\begin{aligned} &1.1\cdot10^6 \le \alpha_1 \le 2.0\cdot10^6 \,(\mathrm{N/m}) \\ &4.0\cdot10^4 \le \alpha_2 \le 5.0\cdot10^4 \,(\mathrm{N/m}) \\ &950 \le \alpha_3 \le 1{,}050\,(\mathrm{kg}) \\ &30 \le \alpha_4 \le 70\,(\mathrm{kg}) \\ &80 \le \alpha_5 \le 120\,(\mathrm{N\cdot s/m}) \end{aligned} \qquad (2.4)$$

There are three functional dependences with five constraints (on the total mass and on the frequencies):

$$\begin{aligned} &f_1(\alpha) = \alpha_3 + \alpha_4 \le 1{,}100.0\,(\mathrm{kg}) \\ &33.0 \le f_2(\alpha) = p_1 = \sqrt{\alpha_1/\alpha_3} \le 42.0\,(s^{-1}) \\ &27.0 \le f_3(\alpha) = p_2 = \sqrt{\alpha_2/\alpha_4} \le 32.0\,(s^{-1}) \end{aligned} \qquad (2.5)$$

The upper limits on the functions f_2 and f_3 are defined approximately and can be significantly modified. In other words, we have two "soft" functional constraints $f_2(\alpha) \le 42.0$, $f_3(\alpha) \le 32.0$; the rest of functional constraints $f_1(\alpha) \le 1{,}100.0$, $33.0 \le f_2(\alpha)$, $27.0 \le f_3(\alpha)$ are rigid. According to Section 2.2, in order to define "soft" constraints on functional relations f_2 and f_3, the latter should be interpreted as the *pseudo-criteria,* that is, $\Phi_1 = f_2$ and $\Phi_2 = f_3$.

The system is to be minimized with respect to the following four performance criteria:

- $\Phi_3 = X_{1\partial}$ (mm): vibration amplitude of the first mass;
- $\Phi_4 = M_1 + M_2$ (kg): metal consumption of the system;

- $\Phi_5 = X_{1\partial} / X_{1st}$: dimensionless dynamical characteristic of the system, where X_{1st} is the static displacement of mass M_1 under the action of the force P;
- $\Phi_6 = \omega / p_1$: dimensionless dynamical characteristic of the system.

Taking the above into account, we formulate the optimization problem. We have a vector of criteria

$$\Phi = (\Phi_1, \Phi_2, \Phi_3, \Phi_4, \Phi_5, \Phi_6)$$

Furthermore, we have design variable constrains (2.4) and three rigid functional constraints

$$1{,}100 \geq f_1(\alpha), 33.0 \leq f_2(\alpha), \text{and}\, 27.0 \leq f_3(\alpha) \tag{2.6}$$

2.4.1 Expert's Difficulties

Because criteria are antagonistic, it is necessary to help the expert define the criteria constraints correctly. As a rule, the expert is ready to change a priori given functional and design variable constraints if it leads to sufficiently improved values of main criteria. Hence, it is necessary to help the expert obtain this information.

Using this example, we will:

1. Show how to define criteria constraints in the interactive mode.
2. Estimate given design variable constraints and show how to correct them.
3. Discuss validity of the given functional constraints.
4. Determine the feasible solution set and the Pareto optimal set.
5. Consider various tools for multicriteria analysis.

The issues 1–4 are particularly relevant because they are always encountered by experts while solving real-life problems.

Thus, the main issue of real-life optimization problem is how to define the feasible solution set. We described a typical statement of optimization problem and the challenge it presents to the expert. By providing this example, we will consider the construction of feasible solutions and the Pareto optimal set when the expert requires help to define the constraints.

In Chapters 3 and 4 we will demonstrate how to correctly state and solve this example using the PSI method and the MOVI software.

References

[1] Statnikov, R. B., and J. B. Matusov, *Multicriteria Analysis in Engineering,* Dordrecht/Boston/London: Kluwer Academic Publishers, 2002.

[2] Statnikov, R. B., and J. B. Matusov, *Multicriteria Optimization and Engineering,* New York: Chapman & Hall, 1995.

[3] Statnikov, R. B., *Multicriteria Design: Optimization and Identification,* Dordrecht/Boston/London: Kluwer Academic Publishers, 1999.

[4] Statnikov, R. B., and J. B. Matusov, "Use of P_τ Nets for the Approximation of the Edgeworth-Pareto Set in Multicriteria Optimization," *Journal of Optimization Theory and Applications,* Vol. 91, No. 3, 1996, pp. 543–560.

[5] Sobol', I. M., and R. B. Statnikov, *Selecting Optimal Parameters in Multicriteria Problems,* 2nd ed., (in Russian), Moscow: Drofa, 2006.

[6] Statnikov, R., et al., "Definition of the Feasible Solution Set in Multicriteria Optimization Problems with Continuous, Discrete, and Mixed Design Variables," *Nonlinear Analysis,* Vol. 71, No. 12, 2009, pp. e109–e117.

[7] Statnikov, R. B., et al., "Multicriteria Analysis Tools in Real-Life Problems," *Journal of Computers and Mathematics with Applications,* Vol. 52, 2006, pp. 1–32.

[8] Statnikov, R. B., A. Bordetsky, and A. Statnikov, "Multicriteria Analysis of Real-Life Engineering Optimization Problems: Statement and Solution," *Nonlinear Analysis,* Vol. 63, 2005, pp. e685–e696.

[9] Statnikov, R. B., A. R. Statnikov, and I. V. Yanushkevich, "Feasible Solutions in Engineering Optimization," *Journal of Machinery Manufacture and Reliability,* Russian Academy of Sciences, No. 4, 2005, pp. 1–9.

[10] Halton, J. H., "On the Efficiency of Certain Quasi-Random Sequences of Points in Evaluating Multi-Dimensional Integrals," *Numerische Mathematik,* Vol. 2, 1960, pp. 84–90.

[11] Hammersley, J. M., "Monte Carlo Methods for Solving Multivariable Problems," *Ann. New York Acad. Sci.,* No. 86, 1960, pp. 844–874.

[12] Hlawka, E., and R. Taschner, *Geometric and Analytic Number Theory,* Springer: Berlin, 1991.

[13] Faure, H., "Discrepancy of Sequences Associated with a Number System (in Dimensions)," *Acta Arithmetica,* Vol. 41, 1982, pp. 337–351, in French.

[14] Kuipers, L., and H. Niederreiter, *Uniform Distribution of Sequences,* New York: John Wiley & Sons, 1974.

[15] Niederreiter, H., "Statistical Independence Properties of Pseudorandom Vectors Produced by Matrix Generators," *J. Comput. and Appl. Math.* No. 31, 1990, pp. 139–151.

[16] Steuer, R. E., and M. Sun, "The Parameter Space Investigation Method of Multiple Objective Nonlinear Programming: A Computational Investigation," *Operations Research,* Vol. 43, No. 4, 2005, pp. 641–648.

3

Using the PSI Method and MOVI Software System for Multicriteria Analysis and Visualization

The PSI method is implemented in the MOVI (Multicriteria Optimization and Vector Identification) software package[1] [1]. Various elements of the MOVI software package allow an expert to: (1) correctly construct the feasible solution set, (2) analyze obtained results, and (3) make a decision about the most preferable solution on the Pareto set. Once the feasible solution set is defined, an expert can use well-known optimization methods for a further search of the optimal solution.

Next we describe visualization tools that are particularly useful for multicriteria analysis. All tools are implemented in the software system MOVI [2–7]:

- Test tables;
- Tables of feasible and Pareto optimal solutions (tables of criteria and design variables);
- Two kinds of histograms: "Design Variable Histograms" (histograms of the distribution of feasible and Pareto optimal solutions) and "Criteria Histograms";
- Two kinds of graphs "Criterion versus Design Variable";
- Graphs "Criterion versus Criterion";
- Tables of functional failures for correcting functional constraints.

1. In this chapter and Chapter 4, all figures were produced using the MOVI software, with the exception of Figure 4.19.

We will give only brief comments on tools since numerous illustrations provided in this chapter are self-explanatory.

Using these techniques developed on the basis of the PSI method, an expert can correctly state and solve the problem under consideration in a series of dialogues with the computer.

3.1 Performing Tests

We consider the example described in Section 2.4. We carry out $N = 1{,}024$ tests (see Figure 3.1). $N1 = 789$ solutions entered test table; these solutions satisfy rigid functional constraints (2.6) and design variable constraints (2.4). Two hundred thirty-five out of 1,024 vectors did not satisfy rigid functional constraints. Generally, it is desirable to check the feasibility of rigid constraints. During the analysis, an expert can revise some of the constraints. In Section 3.4 we will analyze the work of functional constraints.

3.2 Construction of Feasible and Pareto Optimal Sets

3.2.1 Constructing Test Tables

Criteria constraints are determined on the basis of investigation of the test table that is constructed for each criterion. All vectors are arranged in order of decreasing quality. The best solution is located on the top of the test table. For the first criterion, the three best vectors are 480, 608, and 160. For the second criterion, the best vectors are 901, 222, and 476. These solutions are shown and circled in Figure 3.2.

The worst solution is located on the bottom of the test table (see Figure 3.3). The worst vectors for the first criterion are 63, 671, and 959; for the second criterion they are 552, 475, and 889. These solutions are circled; min $\Phi_1^i = 33.01$ corresponds to vector #480 (Figure 3.2), max $\Phi_1^i = 45.60$ corresponds to vector #63 (Figure 3.3), min $\Phi_2^i = 27.00$ corresponds to vector #901 (Figure 3.2), max $\Phi_2^i = 40.64$ corresponds to vector #552 (Figure 3.3).

In Figures 3.2 through 3.6 and later on, the following notations are used:

- N is the number of performed tests.
- $N1$ is the number of tests that entered the test table (that satisfied all design variable and functional constraints).
- NF is the number of feasible solutions (that satisfied all constraints).

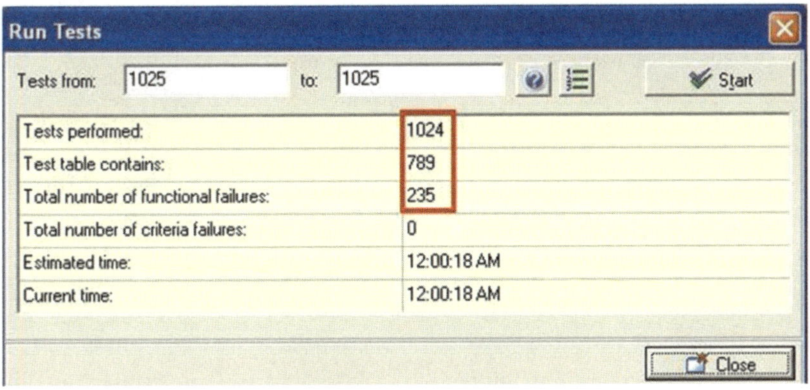

Figure 3.1 Performing tests.

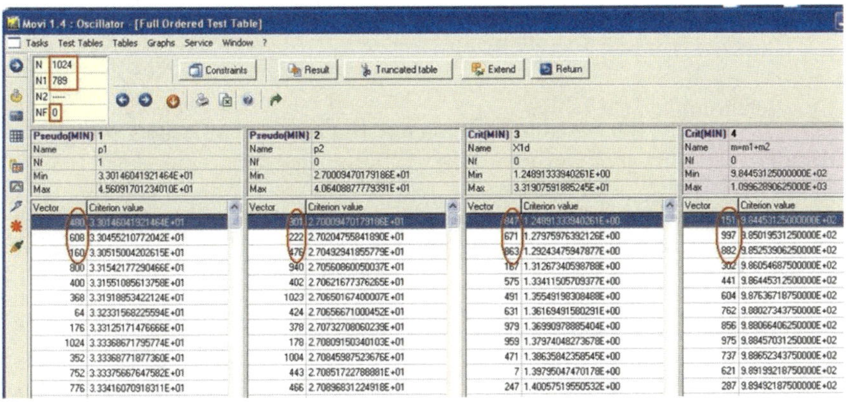

Figure 3.2 The top of the test table. $N = 1{,}024$ is the number of tests. $N1 = 789$ is the number of tests entered the test table. $NF = 0$ is the number of feasible solutions. First dialogue.

- *NP* is the number of Pareto optimal solutions.
- Min and max are the minimum and maximum values of the *i*th criterion in the test table (after *N* tests).
- *Nf* is the number of solutions that satisfied the constraint on the *i*th criterion and all previous criteria constraints simultaneously.

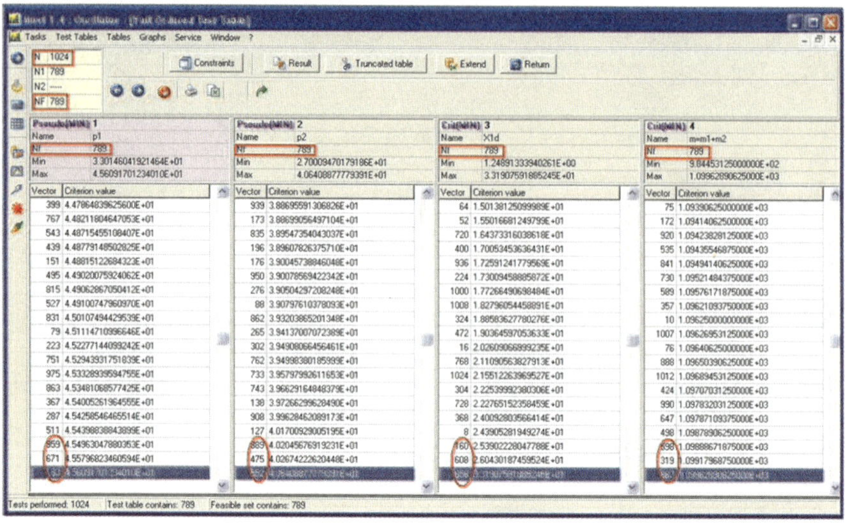

Figure 3.3 The bottom of the test table. Criteria constraints are not imposed, *NF* = 789. The worst solutions are circled.

3.2.2 Constructing the Feasible Solution Set: Dialogues of an Expert with a Computer

First, an expert imposes constraints on each criterion. After that, the computer searches for the common vectors that satisfy all criteria constraints. These vectors are called the *feasible solutions.* If the obtained results are not satisfactory for an expert, he or she revises constraints and/or increases the number of tests.

- *First dialogue* (see Figure 3.2): We have imposed criteria constraints correspondingly to the best solutions by the first criterion (vector 480) and the second criterion (vector 901). As a result, we have obtained an empty feasible solution set, *NF* = 0. What should we do next? We need to revise constraints, for example, relax some of them. Therefore, we make concessions. The case where the criteria constraints have not been imposed at all is shown in Figure 3.3. Therefore, all solutions that satisfy functional constraints are feasible, *NF* = 789.
- *Second dialogue* (see Figure 3.4): By comparison with the first dialogue, we made concessions in all criteria. The first criterion constraint is satisfied by 10 solutions, *Nf* = 10. However, the feasible solution set is empty again, *NF* = 0. We need to make more concessions.
- *Last dialogue* (see Figures 3.5 and 3.6).

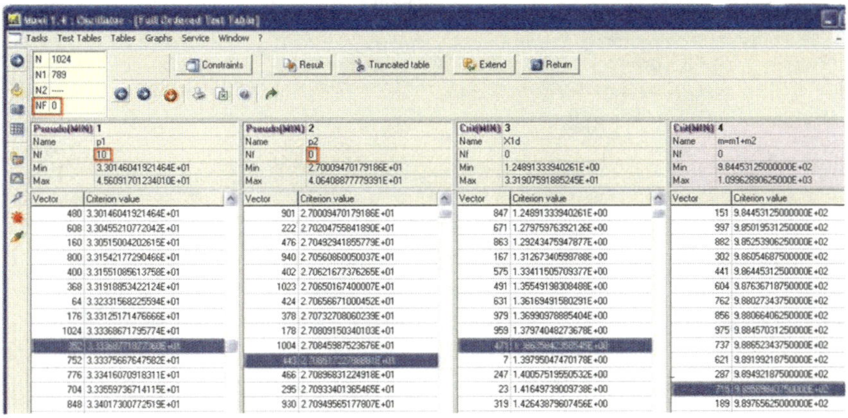

Figure 3.4 Second dialogue.

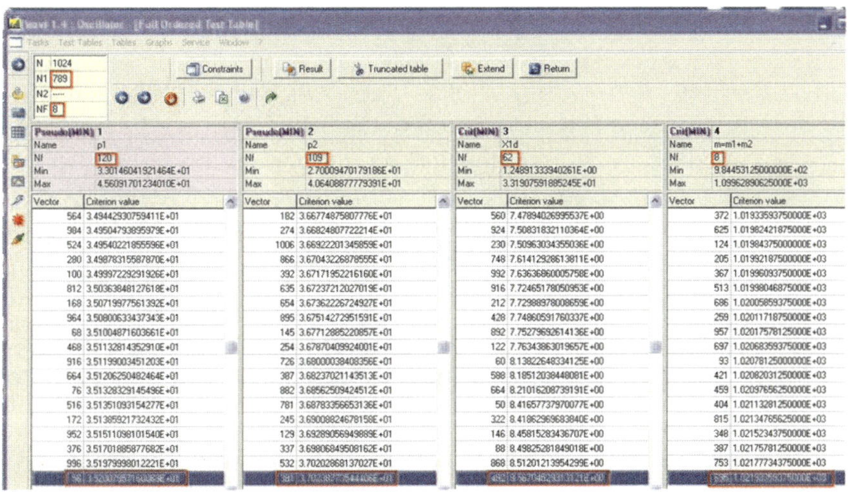

Figure 3.5 Last dialogue. Criteria 1–4.

- *First criterion* (see Figure 3.5): We imposed the constraint $\Phi_1^{**} = 35.200$. This value corresponds to the vector 56. One can see that *Nf* = 120 out of 789 vectors satisfy the first criterion constraint.
- *Second criterion* (see Figure 3.5): We imposed the constraint $\Phi_2^{**} = 37.02$. This value corresponds to the vector 381. *Nf* = 109 vectors satisfy the constraints on the first and second criteria.

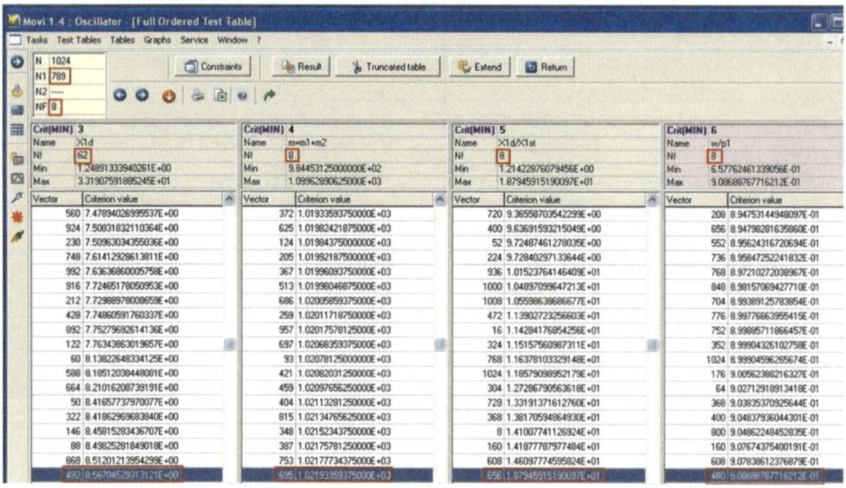

Figure 3.6 Last dialogue. Criteria 3–6.

- *Third criterion* (see Figure 3.5): We imposed the constraint Φ_3^{**} = 8.567. This value corresponds to the vector 492. *Nf*= 62 vectors satisfy the constraints on the first, second, and third criteria.
- *Fourth criterion* (see Figure 3.5): We imposed the constraint Φ_4^{**} = 1,021.93. This value corresponds to the vector 695. *Nf*= 8 vectors satisfy the constraints on the first, second, third, and fourth criteria.
- *Fifth criterion* (see Figure 3.6): We imposed the constraint Φ_5^{**} = 18.794. This value corresponds to the vector 656. *Nf*= 8 vectors satisfy the constraints on the first, second, third, fourth, and fifth criteria.
- *Sixth criterion* (see Figure 3.6): We imposed constraint Φ_6^{**} = 0.908. This value corresponds to the vector 480. *Nf*= 8 vectors satisfy all constraints. A feasible solution set contains eight vectors, *NF*= 8.

The results are shown in Figure 3.7. After 1,024 tests, we obtained eight feasible solutions that satisfy all constraints. Four out of these solutions are Pareto optimal (corresponding to the vectors 520, 336, 672, 288).

3.2.3 Tables of Feasible and Pareto Optimal Solutions

These tables allow an expert to obtain information about the values of criteria and design variables of feasible and Pareto optimal solutions (Figures 3.8 through 3.11).

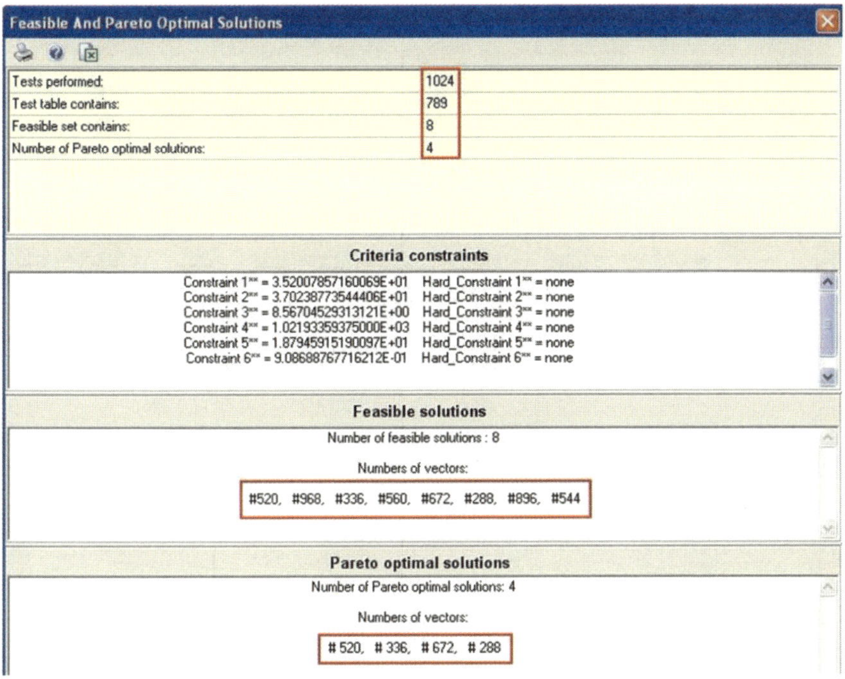

Figure 3.7 Results.

3.2.4 Selecting the Most Preferable Solution

An expert has obtained the Pareto optimal solutions (see Figures 3.7 and 3.8). Then he or she needs to analyze the Pareto optimal solutions (see Figures 3.8 and 3.10) and define the best solution. As indicated in the Section 1.1, the analysis of the Pareto optimal set to determine the most preferable solution does not pose a challenge to an expert in our type of problems. This is because experts have a sufficiently well-defined system of preferences and the Pareto optimal set often contains only a few solutions due to stringent constraints.

In our case we have two pseudo-criteria (Criteria 1 and 2) and four performance criteria (Criteria 3–6). We have obtained four Pareto optimal solutions 520, 336, 672, and 288 (see Figures 3.7 and 3.8).

It is follows from Figure 3.8:

- The minimum value of Criterion 3 (equal to 2.932) corresponds to solution 288.
- The minimum value of Criterion 4 (equal to 1,002.77) corresponds to solution 336.

Table of Criteria

Criterion: Pareto optimal set ☐ Show pseudocriteria

Tests performed: 1024 Feasible set contains: 8 Pareto optimal set contains: 4

Vector	3 - X1d -> MIN	4 - m=m1+m2 -> MIN	5 - X1d/X1st -> MIN	6 - w/p1 -> MIN
Min	2.93217659307383E+00	1.00277343750000E+03	1.63589110119441E+00	8.60342400146053E-01
Max	3.93462597292253E+00	1.01630859375000E+03	2.27643472427534E+00	8.80472559017949E-01
288	2.93217659307383E+00	1.00980468750000E+03	1.63589110119441E+00	8.80472559017949E-01
336	3.69972189596129E+00	1.00277343750000E+03	2.10313292542878E+00	8.68358046324039E-01
520	3.93462597292253E+00	1.00873046875000E+03	2.27643472427534E+00	8.60342400146053E-01
672	2.99734527150947E+00	1.01630859375000E+03	1.67620094700186E+00	8.77062996920710E-01

Figure 3.8 Table of criteria: Pareto optimal solutions.

Table of Criteria

Criterion: Feasible set ☐ Show pseudocriteria

Tests performed: 1024 Feasible set contains: 8 Pareto optimal set contains: 4

Vector	1 - p1 -> MIN	2 - p2 -> MIN	3 - X1d -> MIN	4 - m=m1+m2 -> MIN	5 - X1d/X1st -> MIN	6 - w/p1 -> MIN
Min	3.38144218150189E+01	2.70971835801064E+01	2.93217659307383E+00	1.00277343750000E+03	1.63589110119441E+00	8.60342400146053E-01
Max	3.48698378632823E+01	3.11909661071071E+01	7.47894026995537E+00	1.01779296875000E+03	4.27446268846814E+00	8.87195415143112E-01
288	3.40726121362158E+01	2.97191238807196E+01	2.93217659307383E+00	1.00980468750000E+03	1.63589110119441E+00	8.80472559017949E-01
336	3.45479610939254E+01	3.03747971980251E+01	3.69972189596129E+00	1.00277343750000E+03	2.10313292542878E+00	8.68358046324039E-01
520	3.48698378632823E+01	2.70971835801064E+01	3.93462597292253E+00	1.00873046875000E+03	2.27643472427534E+00	8.60342400146053E-01
544	3.38144218150189E+01	3.07983138997379E+01	5.91774890807082E+00	1.01779296875000E+03	3.29897164469944E+00	8.87195415143112E-01
560	3.44040460714369E+01	3.11909661071071E+01	7.47894026995537E+00	1.01216796875000E+03	4.27446268846814E+00	8.71990461171565E-01
672	3.42050686271423E+01	2.86452467577484E+01	2.99734527150947E+00	1.01630859375000E+03	1.67620094700186E+00	8.77062996920710E-01
896	3.39851634205505E+01	2.85109847985053E+01	3.09703302724984E+00	1.01611328125000E+03	1.71289517088178E+00	8.82738141605030E-01
968	3.48161919715982E+01	3.09184355482384E+01	5.29991285962000E+00	1.01595703125000E+03	3.09894807099363E+00	8.61668042974744E-01

Figure 3.9 Table of criteria: feasible solutions.

Design Variable Values

Vectors from 1 to 2 Pareto optimal set Show ☐ Hide Constant Variables

Tests performed 1024 Feasible set contains: 8 Pareto optimal set contains: 4

Number of vector	Variable # 1 - k1	Variable # 2 - k2	Variable # 3 - m1	Variable # 4 - m2	Variable # 5 - c
Min	1.11582031250000E+06	4.19042968750000E+04	9.51660156250000E+02	4.86718750000000E+01	1.02148437500000E+02
Max	1.15712890625000E+06	4.95214843750000E+04	9.61132812500000E+02	6.03515625000000E+01	1.19492187500000E+02
520	1.15712890625000E+06	4.19042968750000E+04	9.51660156250000E+02	5.70703125000000E+01	1.02148437500000E+02
336	1.13691406250000E+06	4.63476562500000E+04	9.52539062500000E+02	5.02343750000000E+01	1.14140625000000E+02
672	1.11845703125000E+06	4.95214843750000E+04	9.55957031250000E+02	6.03515625000000E+01	1.19492187500000E+02
288	1.11582031250000E+06	4.29882812500000E+04	9.61132812500000E+02	4.86718750000000E+01	1.03203125000000E+02

Figure 3.10 Table of design variables: Pareto optimal solutions.

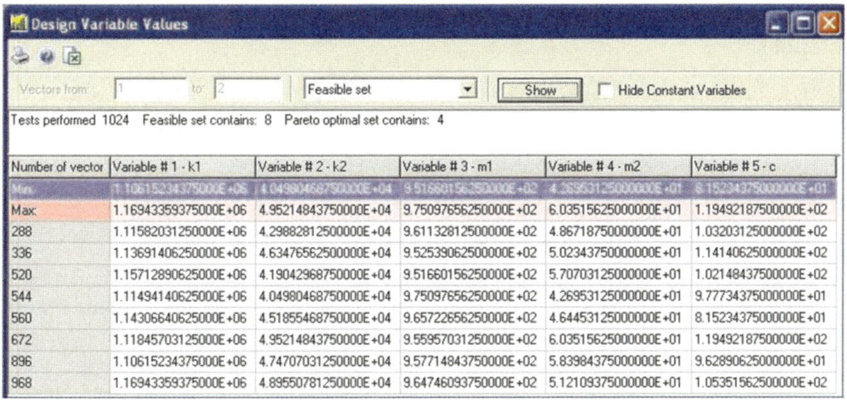

Number of vector	Variable # 1 - k1	Variable # 2 - k2	Variable # 3 - m1	Variable # 4 - m2	Variable # 5 - c
Min:	1.10615234375000E+06	4.04980468750000E+04	9.51660156250000E+02	4.26953125000000E+01	8.15234375000000E+01
Max:	1.16943359375000E+06	4.95214843750000E+04	9.75097656250000E+02	6.03515625000000E+01	1.19492187500000E+02
288	1.11582031250000E+06	4.29882812500000E+04	9.61132812500000E+02	4.86718750000000E+01	1.03203125000000E+02
336	1.13691406250000E+06	4.63476562500000E+04	9.52539062500000E+02	5.02343750000000E+01	1.14140625000000E+02
520	1.15712890625000E+06	4.19042968750000E+04	9.51660156250000E+02	5.70703125000000E+01	1.02148437500000E+02
544	1.11494140625000E+06	4.04980468750000E+04	9.75097656250000E+02	4.26953125000000E+01	9.77734375000000E+01
560	1.14306640625000E+06	4.51855468750000E+04	9.65722656250000E+02	4.64453125000000E+01	8.15234375000000E+01
672	1.11845703125000E+06	4.95214843750000E+04	9.55957031250000E+02	6.03515625000000E+01	1.19492187500000E+02
896	1.10615234375000E+06	4.74707031250000E+04	9.57714843750000E+02	5.83984375000000E+01	9.62890625000000E+01
968	1.16943359375000E+06	4.89550781250000E+04	9.64746093750000E+02	5.12109375000000E+01	1.05351562500000E+02

Figure 3.11 Table of design variables: feasible solutions.

- The minimum value of Criterion 5 (equal to 1.636) corresponds to solution 288.
- The minimum value of Criterion 6 (equal to 0.86) corresponds to solution 520.

In Figure 3.8 these solutions are circled.

An expert may prefer solution 336 with a minimum of metal consumption, or solution 288 with the best values of the amplitude of the first mass and dynamical characteristic X_{1d}/X_{1st}, or solution 520 with a minimum of the dynamical characteristic ω/p_1, or a compromise solution 672.

3.3 Histograms and Graphs

So an expert has defined the feasible and Pareto optimal sets. Several important questions arise:

- How reasonable were the design variable constraints given in the initial problem?
- Is it possible to improve obtained Pareto optimal solutions?

Analysis of test tables and tables of feasible and Pareto optimal solutions, histograms, and graphs allows us to answer these questions.

3.3.1 Design Variable Histograms: Histograms of the Distribution of Feasible Solutions

We have previously obtained the tables of feasible solutions (see Figure 3.11). Taking into account this table, we now construct the histograms. The interval of variation of the design variable $[\alpha_j^*;\alpha_j^{**}]$, $j = 1, \ldots, r$ is divided into 10 identical subintervals (see Figures 3.12 through 3.16). Above each subinterval, the number of feasible solutions entering this subinterval is indicated. Analyzing the histograms as well as tables of feasible solutions reveals how these solutions are distributed in design variable space. The histograms and tables of feasible solutions play an important role in correcting the design variable and other constraints.

The regions of the feasible and Pareto optimal solutions are circled in Figures 3.12, 3.14, and 3.15. All eight feasible solutions are located on the left end of the range of change for the first and third design variables (Figures 3.12 and 3.14). In these cases it can be recommended to decrease the values of the left and right borders for these design variables. The feasible solutions for the fourth design variable are located in the middle of the interval (Figure 3.15). Here the value of the left border can be increased and the right border decreased.

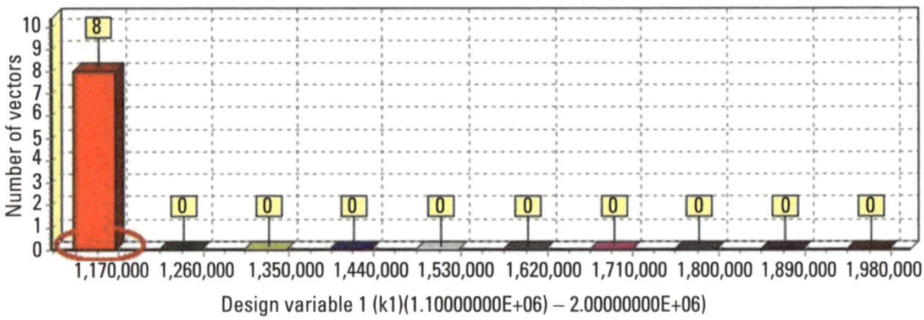

Figure 3.12 First design variable.

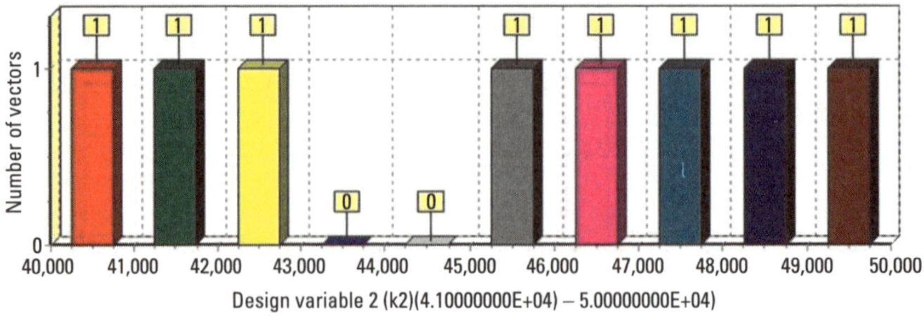

Figure 3.13 Second design variable.

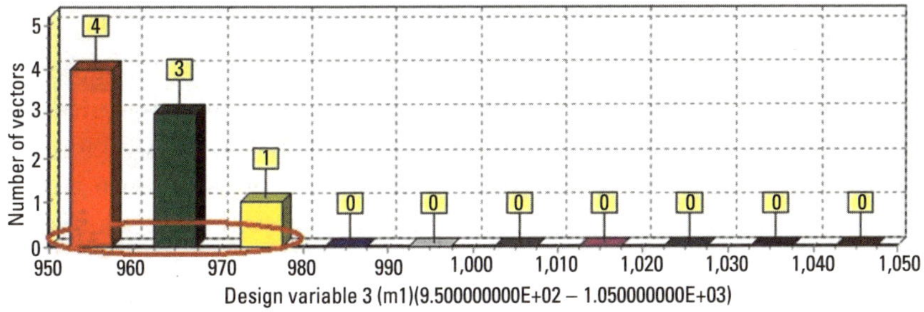

Figure 3.14 Third design variable.

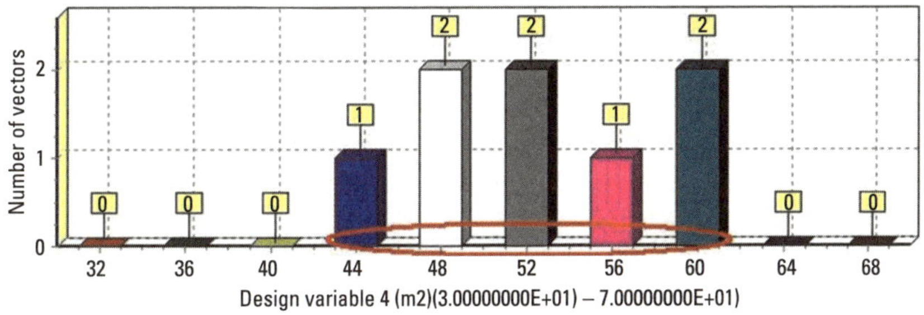

Figure 3.15 Fourth design variable.

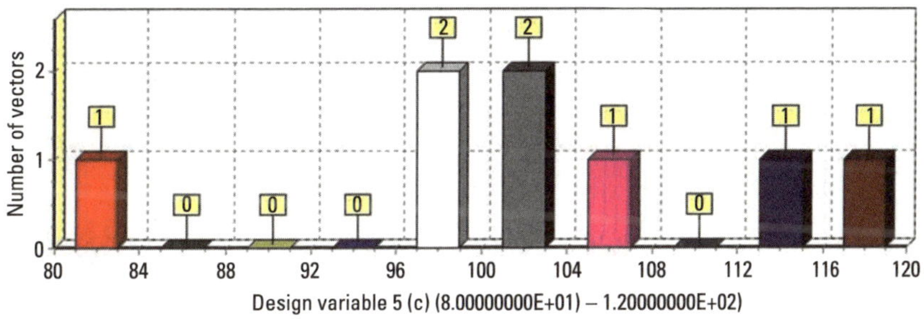

Figure 3.16 Fifth design variable.

On the other hand, the feasible solutions for the second and the fifth design variables are more or less uniformly distributed along the interval.

The analysis of the histograms allows us to see the work of constraints and is helpful for the correction of the initial design variable constraints.

3.3.2 Criteria Histograms: Visualization of Contradictory Criteria

Criteria histograms are constructed on the basis of criteria tables (Figure 3.9). The next criteria histograms are given taking into account pseudo-criteria and criteria (Figures 3.17 through 3.22). Eight feasible solutions, #288, #336, #520, #544, #560, #672, #896, and #968, are located in corresponding intervals: [33.81; 34.87] for the first pseudo-criterion; [27.1; 31.19] for the second pseudo-criterion; [2.93; 7.47] for the third criterion; [1,003; 1,018] for the fourth criterion; [1.63; 4.27] for the fifth criterion; and [0.860; 0.887] for the sixth criterion. The feasible solutions are shown with color bars on the histograms.

Using our simple mathematical model (2.3), (2.4), and (2.6), we show that the relations between criteria are very complex. This situation is typical to real-life problems. Therefore, finding all the feasible and Pareto optimal solutions is an important problem for correct decision-making. On the basis of the PSI method and multicriteria analysis, an expert obtains this relevant information.

3.3.3 Graphs "Criterion Versus Design Variable II"

We consider projections of the *multidimensional points* $\Phi_v(\alpha^i)$, $v = 1, \ldots, k$, $i = 1, \ldots, N1$ onto the plane $\Phi_v \alpha_j$. These projections provide information about

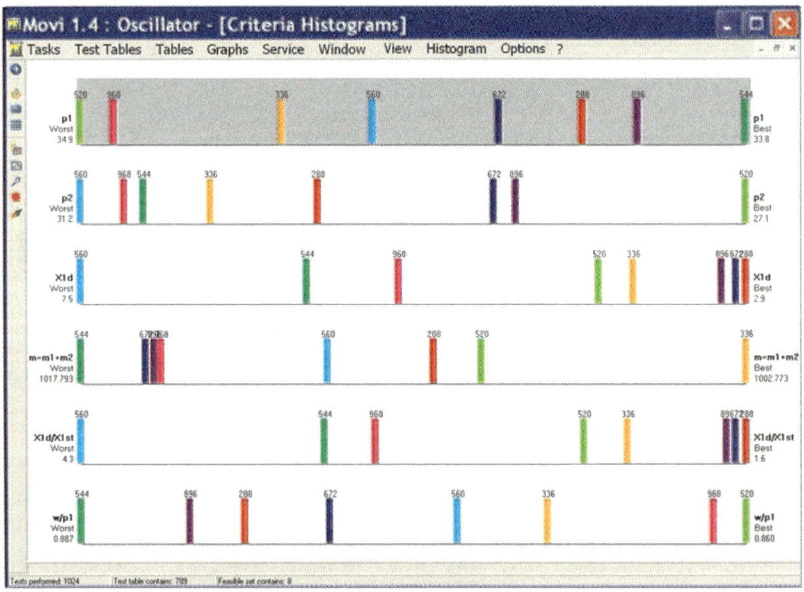

Figure 3.17 Histograms of feasible solutions.

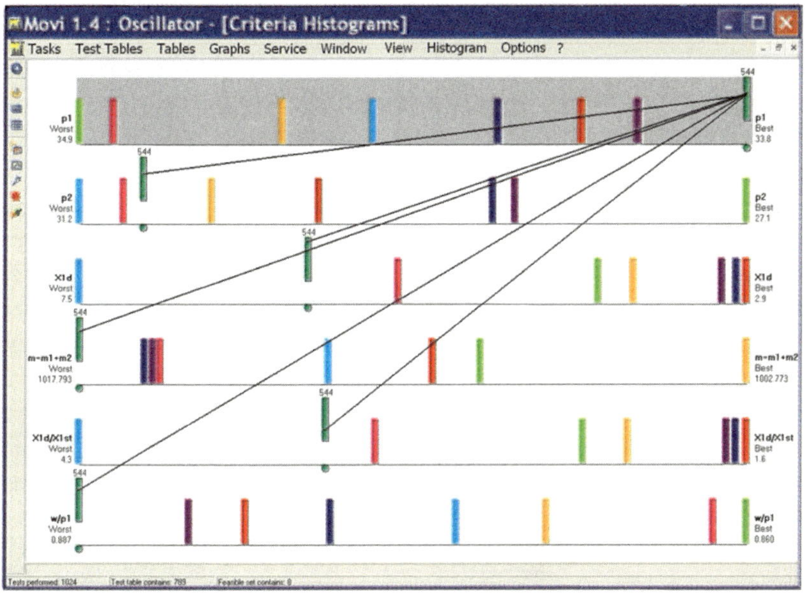

Figure 3.18 Solution #544 is the best by the first pseudo-criterion and the worst by the fourth and sixth criteria.

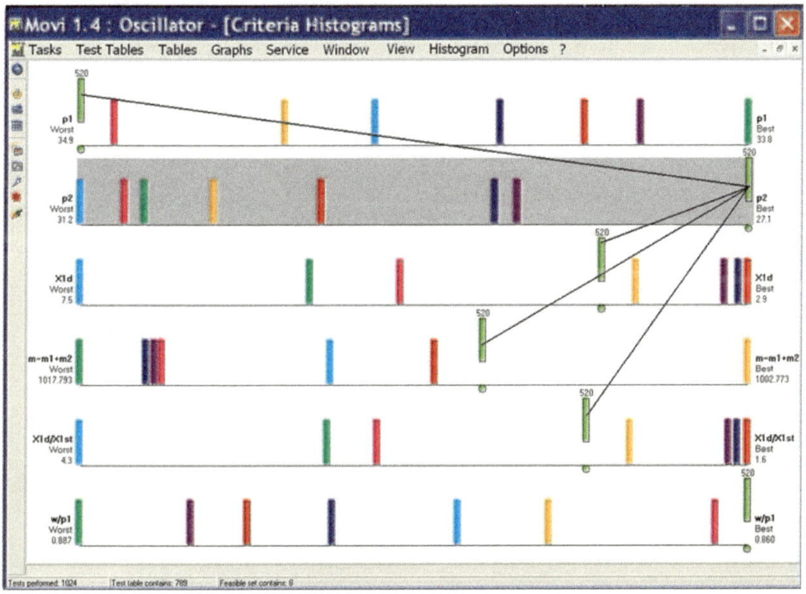

Figure 3.19 The Pareto optimal solution #520 is the best by the sixth criterion and the second pseudo-criterion, and the worst by the first pseudo-criterion.

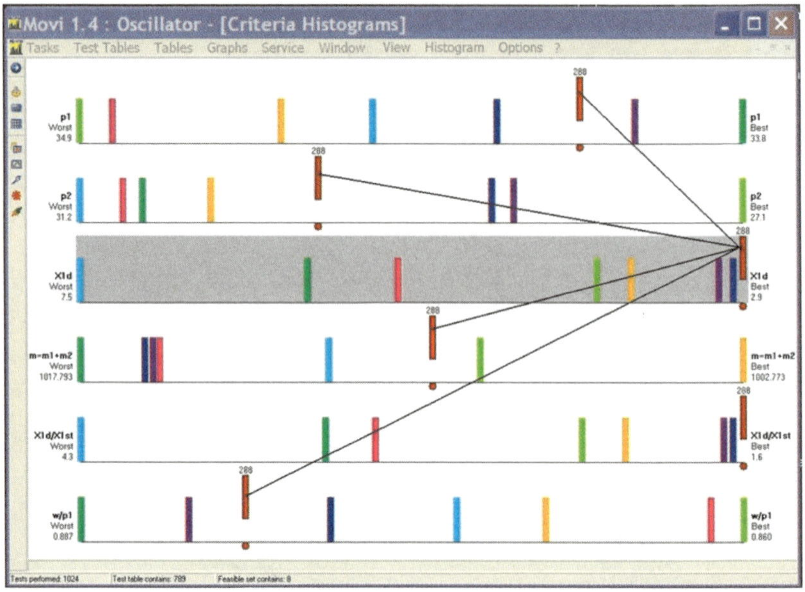

Figure 3.20 The Pareto optimal solution #288 is the best by the third and fifth criteria.

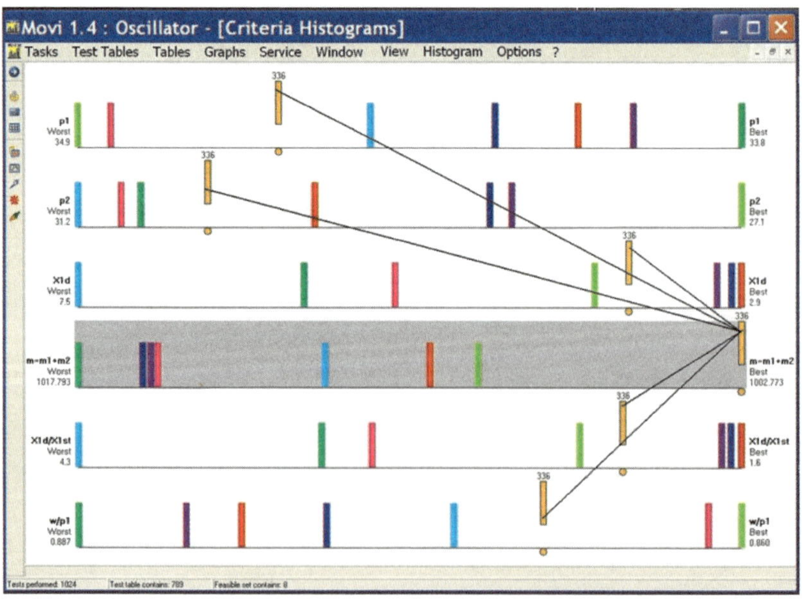

Figure 3.21 The Pareto optimal solution #336 is the best by the fourth criterion.

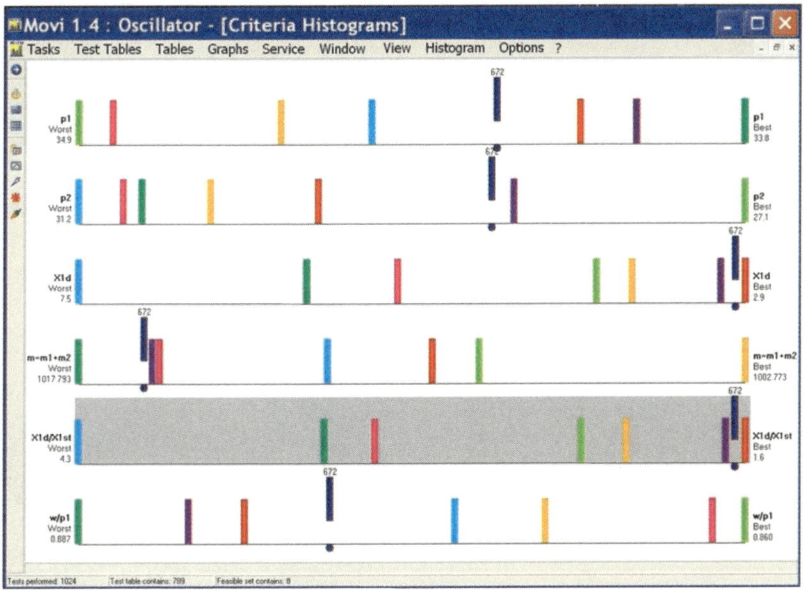

Figure 3.22 The Pareto optimal solution #672.

the sensitivity of criteria to the design variables and also point to localization of the feasible solution set.

In our case $N1 = 789$ solutions have entered the test table after $N = 1{,}024$ tests (Figures 3.1 and 3.7), $k = 6$, $\alpha^i = \alpha_1^i, \ldots, \alpha_5^i$. Graphs "Criterion versus Design Variable II" are shown in Figures 3.23 through 3.26. We have obtained eight feasible solutions (blue and green points) and four Pareto optimal solutions (green points). Unfeasible solutions are shown with magenta. One can see that the first and sixth criteria are antagonistic with respect to the first design variable and strongly depend on this parameter (Figures 3.23 and 3.26). It is not obvious whether the second criterion depends on the first design variable. The third criterion weakly depends on the first design variable.

3.3.4 Graphs "Criterion Versus Criterion"

After N tests, $N1$ design variable vectors have entered the test table. We consider projections of the *multidimensional points* $\Phi_v(\alpha^i)$, $v = 1, \ldots, k$, $i = 1, \ldots, N1$ onto the plane $\Phi_i\Phi_j$. These projections provide information about dependences between criteria and localization of the feasible solution and Pareto optimal sets in the criteria space. This information helps to improve the statement and solution of optimization problem and finally to estimate a correctness of the mathematical model and its shortcomings.

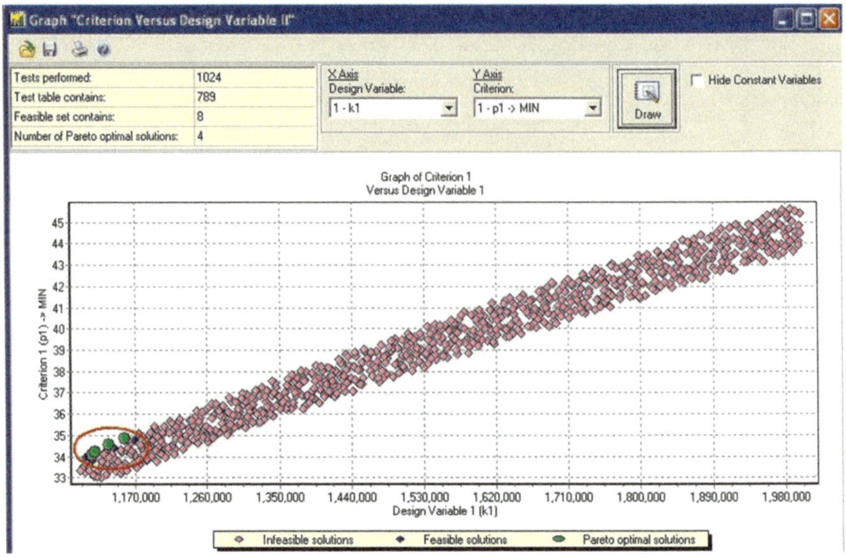

Figure 3.23 First criterion versus first design variable. The area of feasible solutions (blue and green points) including Pareto optimal solutions (green points) is circled.

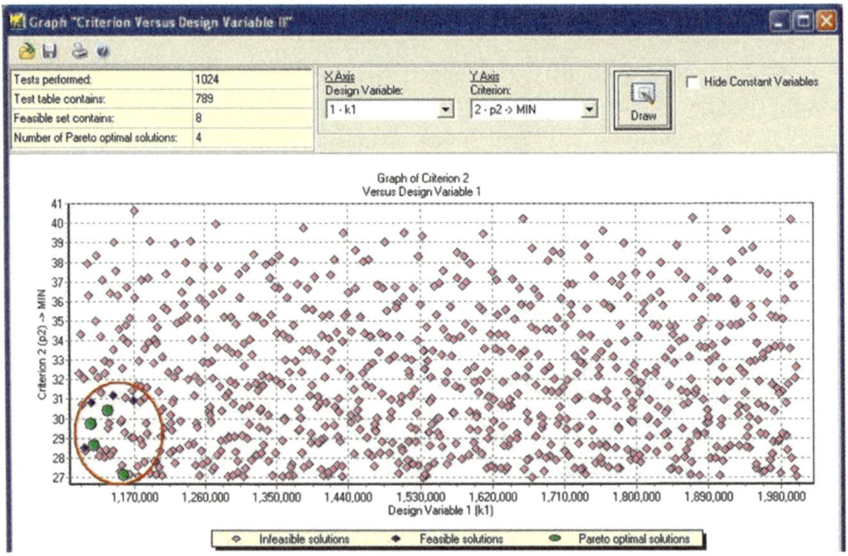

Figure 3.24 Second criterion versus first design variable. The area of feasible solutions (blue and green points) including Pareto optimal solutions (green points) is circled.

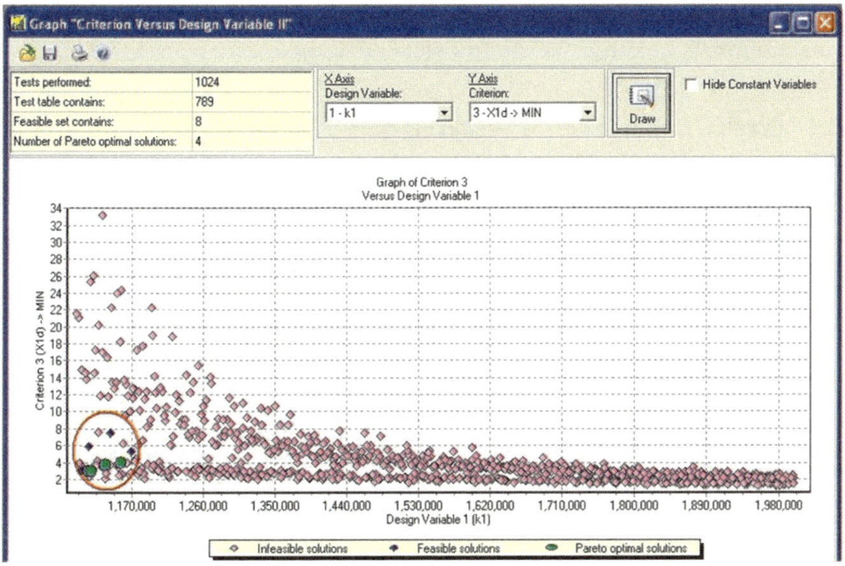

Figure 3.25 Third criterion versus first design variable. The area of feasible solutions (blue and green points) including Pareto optimal solutions (green points) is circled.

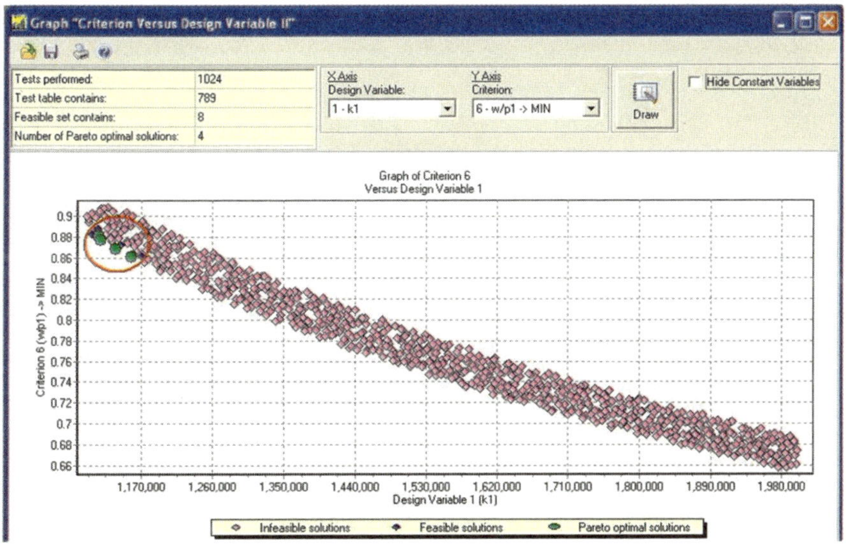

Figure 3.26 Sixth criterion versus first design variable. The area of feasible solutions (blue and green points) including Pareto optimal solutions (green points) is circled.

In Figures 3.27 through 3.31, one can see projections of multidimensional points $\Phi_v(\alpha^i)$ onto the criteria plane $\Phi_i\Phi_j$.

3.3.5 Graphs "Criterion Versus Design Variable I"

Consider that after the analysis of the test table, the preference is given to the Pareto optimal solution α^i. Let us fix all components of this vector, apart from the *j*th one, α^i_j, and find out how functions $\Phi_1, \ldots, \Phi_k$ (or some of them) change as the component α^i_j varies on the original interval, $[\alpha^*_j;\alpha^{**}_j]$, The sensitivity of criteria to parameters for the Pareto optimal solution 288 (which is one of the four solutions that were found previously) is illustrated in Figures 3.32 through 3.34. In Figures 3.32 and 3.33 we investigated the influence of the third design variable on the first and sixth criteria. All the other design variables of the Pareto optimal solution 288, except the third design variable, were unchanged. One can see that the first and sixth criteria are antagonistic with respect to the third design variable and strongly depend on this parameter. Complex dependences between the fifth criterion and fourth design variable are shown in the Figure 3.34. In this case all design variables were fixed, except for the fourth design variable. In Figures 3.32, 3.33, and 3.34, the Pareto optimal solution 288 plays the role of the prototype; the prototype is denoted as a red circle with a yellow filling.

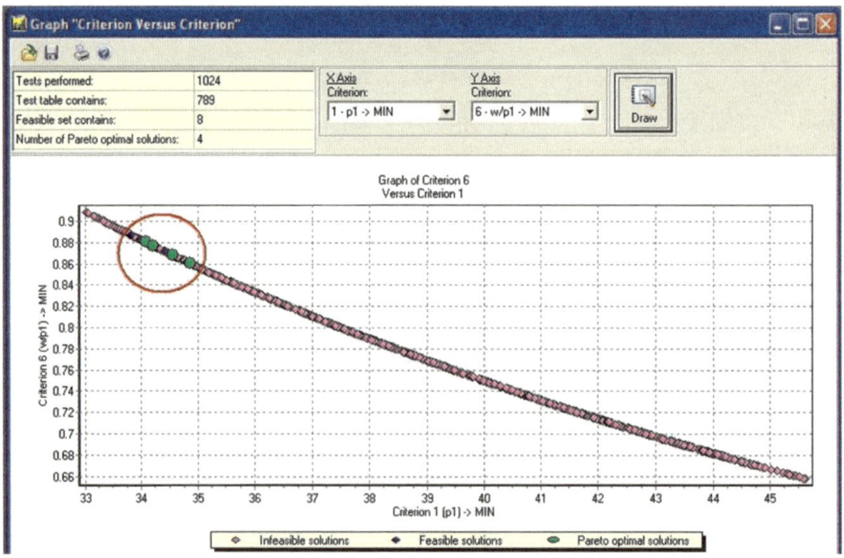

Figure 3.27 First criterion versus sixth criterion. The area of feasible solutions (blue and green points) including Pareto optimal solutions (green points) is circled.

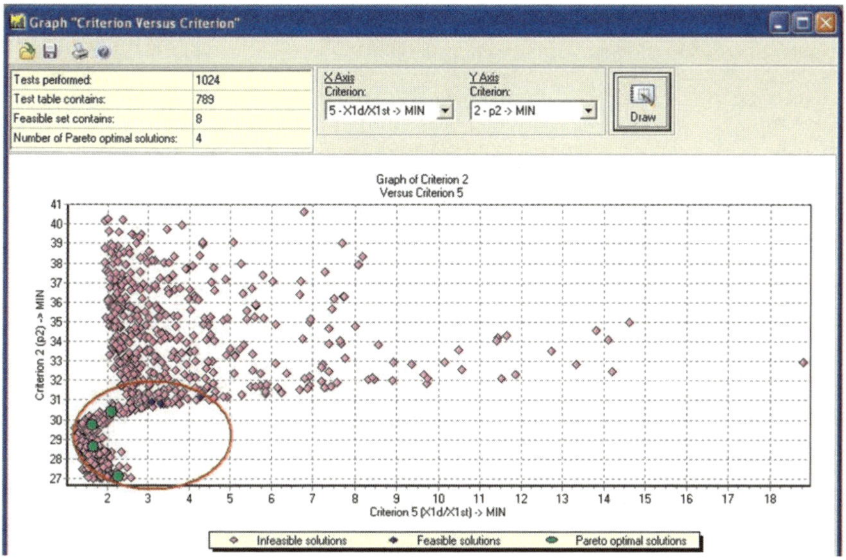

Figure 3.28 Fifth criterion versus second criterion. The area of feasible solutions (blue and green points) including Pareto optimal solutions (green points) is circled.

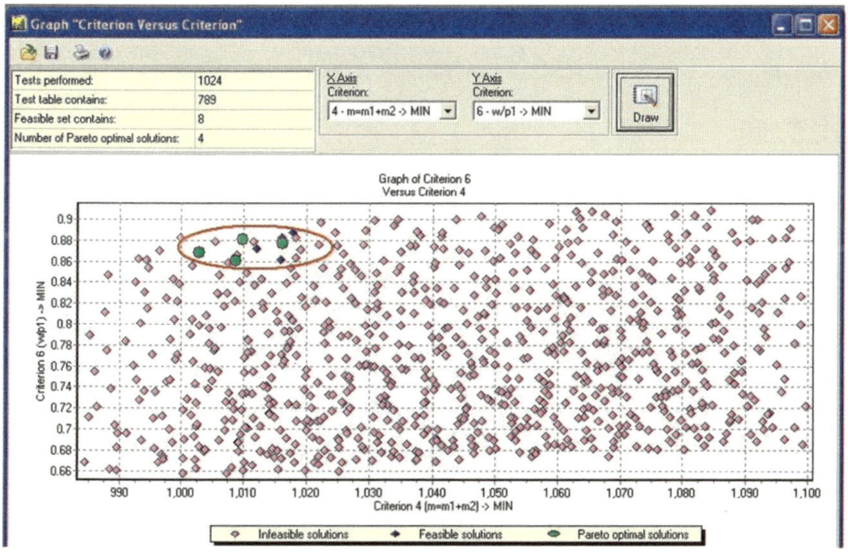

Figure 3.29 Fourth criterion versus sixth criterion. The area of feasible solutions (blue and green points) including Pareto optimal solutions (green points) is circled.

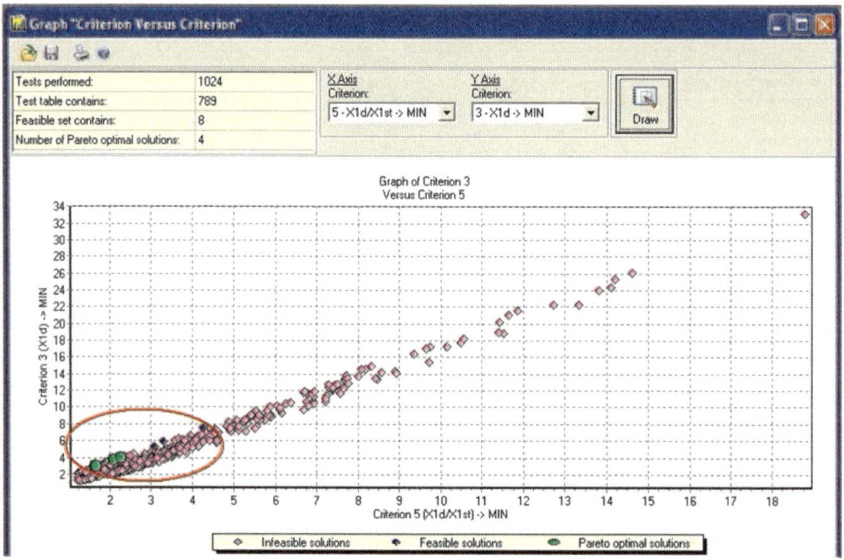

Figure 3.30 Third criterion versus fifth criterion. The area of feasible solutions (blue and green points) including Pareto optimal solutions (green points) is circled.

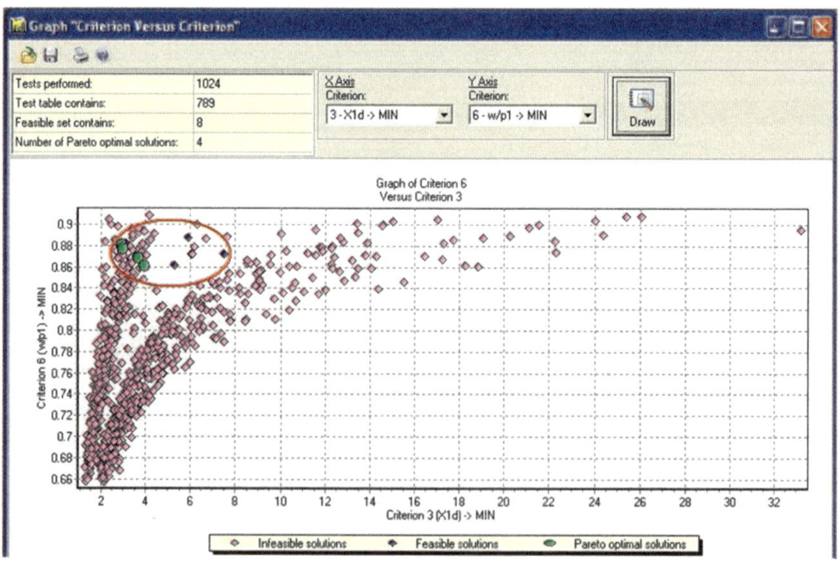

Figure 3.31 Third criterion versus sixth criterion. The area of feasible solutions (blue and green points) including Pareto optimal solutions (green points) is circled.

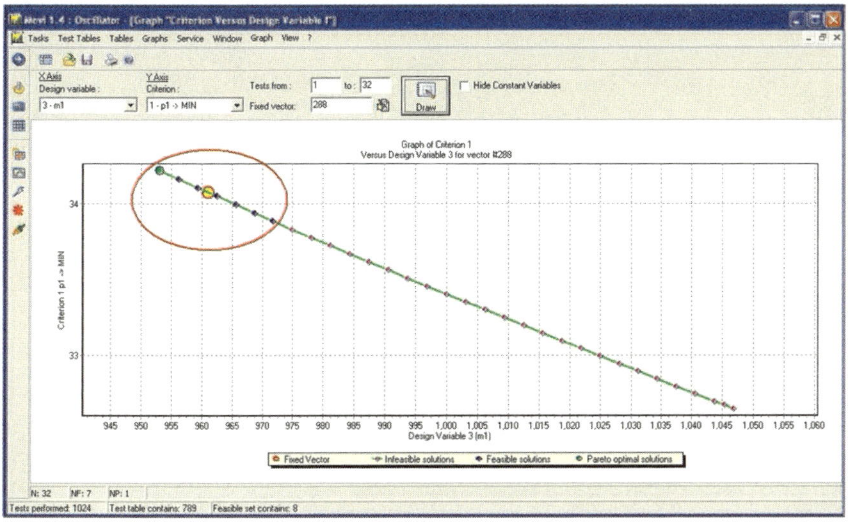

Figure 3.32 First criterion versus third design variable, $N = 32$, $NF = 7$, $NP = 1$. The area of feasible solutions (blue and green points) including Pareto optimal solutions (green points) is circled.

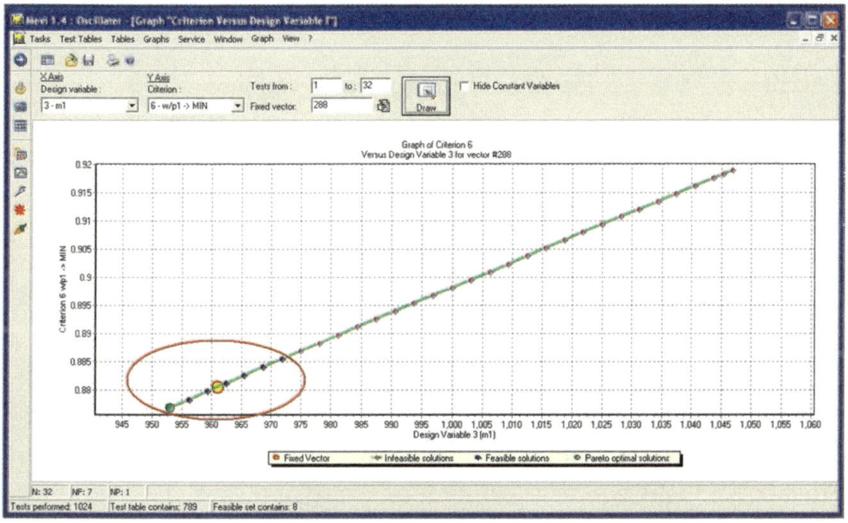

Figure 3.33 Sixth criterion versus third design variable, $N = 32$, $NF = 7$, $NP = 1$. The area of feasible solutions (blue and green points) including Pareto optimal solutions (green points) is circled.

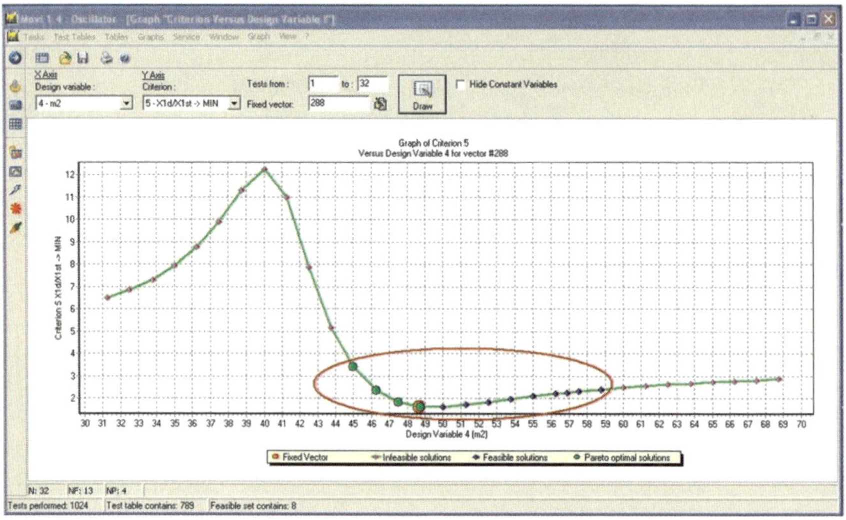

Figure 3.34 Fifth criterion versus fourth design variable, $N=32$, $NF=13$, $NP=4$. The area of feasible solutions (blue and green points) including Pareto optimal solutions (green points) is circled.

3.4 Weakening Functional Constraints

In Section 2.2 we considered an approach for determining "soft" functional constraints. In order to do this, we presented functional dependences as pseudo-criteria. In this section, we will discuss another approach used to revise functional constraints. This approach is based on tables of functional failures [1]. It allows us to assess how reasonable the constraints are and how they actually work. This problem is especially important if the feasible solution set is very poor or empty. In our case an expert can weaken functional constraints based on the analysis of functional failures tables.

3.4.1 Tables of Functional Failures

Recall that only those solutions that satisfy all functional constraints enter the test table. Solutions that fail to satisfy these constraints enter the table of functional failures [1]. This table contains information about all the design variable vectors that do not satisfy the functional constraints; information includes the total number of tests conducted, the total number of functional failures, and the number of functional failures for each functional constraint c_j.

Earlier we solved our problem with three rigid functional constraints (2.6) and obtained only eight feasible solutions after 1,024 tests (see Section 3.2.2 and Figure 3.7). Next we show how to correct these constraints using tables of functional failures. Consider the table of functional failures for the first functional constraint (see Figure 3.35).

We see that after 1,024 tests, 235 solutions did not satisfy all functional constraints:

- Fifty-two solutions did not satisfy the first functional constraint.
- Seven solutions did not satisfy the second functional constraint.
- One hundred seventy-six solutions did not satisfy the last functional constraint. This constraint rejected the largest number of solutions.

The upper and lower parts of the table of functional failures for the first functional constraint are shown in Figure 3.35. These are 52 unfeasible vectors 162, 976, 53, ..., 157, 860, and 479 that did not satisfy the first criterion

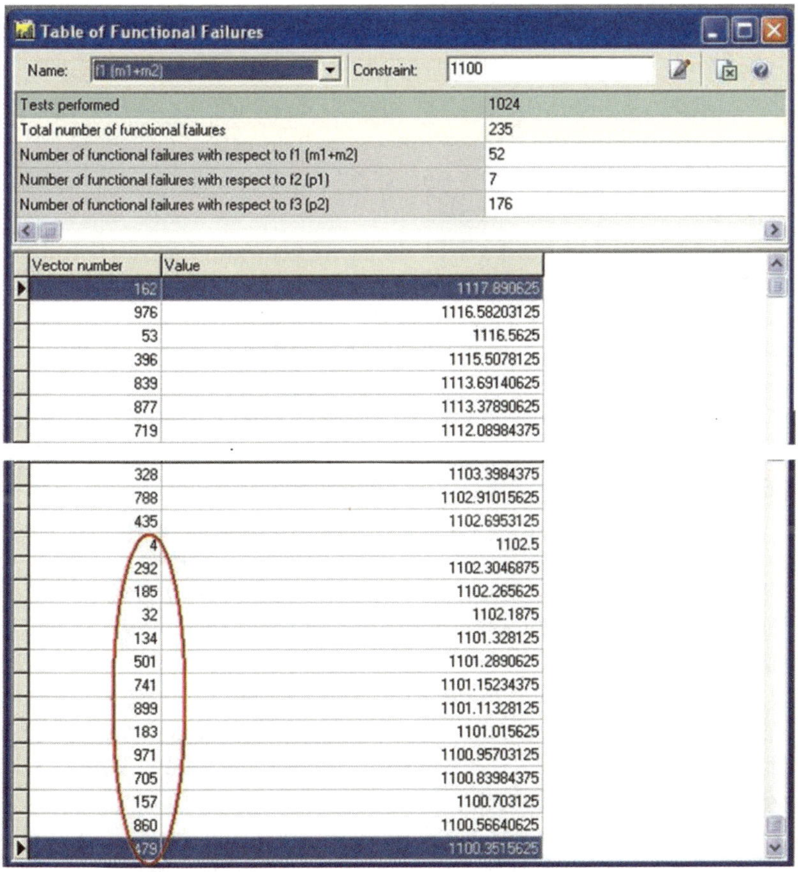

Tests performed	1024
Total number of functional failures	235
Number of functional failures with respect to f1 (m1+m2)	52
Number of functional failures with respect to f2 (p1)	7
Number of functional failures with respect to f3 (p2)	176

Vector number	Value
162	1117.890625
976	1116.58203125
53	1116.5625
396	1115.5078125
839	1113.69140625
877	1113.37890625
719	1112.08984375
328	1103.3984375
788	1102.91015625
435	1102.6953125
4	1102.5
292	1102.3046875
185	1102.265625
32	1102.1875
134	1101.328125
501	1101.2890625
741	1101.15234375
899	1101.11328125
183	1101.015625
971	1100.95703125
705	1100.83984375
157	1100.703125
860	1100.56640625
479	1100.3515625

Figure 3.35 Table of functional failures for the first functional constraint (fragment).

constraint, $f_1 \leq 1{,}100$. All vectors are arranged in decreasing order of the value f_1. So the "worst" vector is 162 (f_1 equal to 1,117.89) and the "best" vector is 479 (f_1 equal to 1,100.35).

The obtained information allows an expert to estimate the values assumed by the design variable vectors that do not satisfy the *j*th functional constraint and to find out whether it is advisable to change this constraint so that as many solutions as possible will satisfy it.

To do this, an expert selects the functional dependency f_j. Then, using the table, an expert determines the values assumed by the function f_j and estimates to what extent one can change the corresponding value c_j so that a number of solutions will meet this constraint. Next, one introduces a new (relaxed) value c_j. The solutions that satisfy this new constraint will then be tested to find out if they satisfy all subsequent constraints, if the latest exist. The solutions that satisfy all functional constraints enter the test table. Taking into account a possible shortage of computer time for calculating the functional dependences, it is advisable to relax the functional constraints, beginning with the last one, c_p. Then, if necessary and possible, the constraint c_{p-1} can be relaxed, and so on. When relaxing functional constraints, an expert can also proceed as follows. First, one identifies all functional dependences on which the constraints can be relaxed and then, using the table of functional failures, defines a new (relaxed) value of the functional constraint and enters this change in the tables of functional failures.

Using the tables of functional failures, an expert obtains all the information required to correct the functional constraints. The analysis of the tables of functional failures generally does not cause any difficulties.

One hundred seventy-six vectors do not satisfy the *third functional constraint*, $27.0 \leq f_3(\alpha)$ (see Figure 3.36). It is evident from Figure 3.36 that the "best" vector is 278 (f_3 equal to 26.98). If we relax this constraint a little, for example, from 27 to 26.835, 12 vectors (278, 551, 833, 907, 545, 452, 81, 948, 649, 507, 973, and 112) will not only satisfy the new constraint but also enter the test table. This is because this constraint is the last one. These vectors are circled in the figure.

Seven vectors do not satisfy the *second functional constraint,* $33.0 \leq f_2(\alpha)$ (see Figure 3.37). One can see from Figure 3.37 that the "best" vector is 384 (f_2 equal to 32.99), and the "worst" vector is 640 (f_2 equal to 32.46). If we relax this constraint a little, for example, from 33 to 32.85, three vectors (384, 80, and 192) will satisfy the new constraint. These vectors are circled in the figure. Furthermore, solutions 384 and 80 will enter the test table, because $27.0 \leq f_3(384) = 32.99$ and $27.0 \leq f_3(80) = 30.625$.

If we relax the *first functional constraint* a little (e.g., from 1,100 to 1,102.6), 14 vectors (479, 860, 157, 705, 971, 183, 899, 741, 501, 134, 32, 185, 292, and 4) will satisfy the new constraint (see Figure 3.35). These vectors

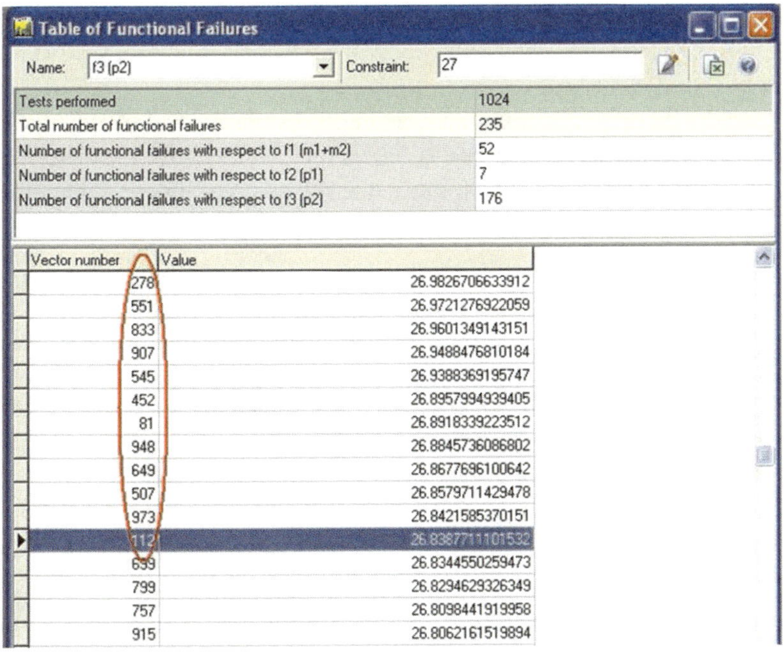

Figure 3.36 Table of functional failures for the third functional constraint (fragment).

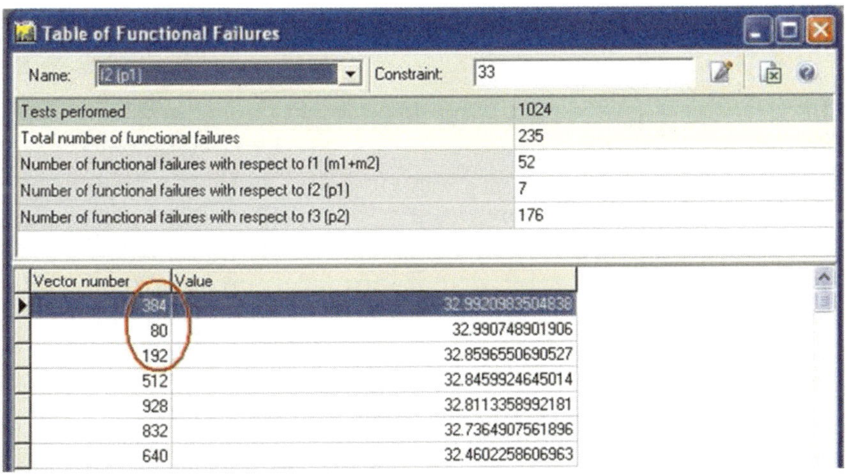

Figure 3.37 Table of functional failures for the second functional constraint.

are circled in the figure. Furthermore, nine solutions (157, 705, 971, 183, 899, 741, 134, 185, and 292) will enter the test table, because these vectors satisfy constraints $33.0 \leq f_2(\alpha)$ and $27.0 \leq f_3(\alpha)$.

Unjustifiably rigid functional constraints can lead to a loss of interesting solutions and sometimes to an empty feasible solution set. We have shown how one can correct functional constraints using the tables of functional failures.

References

[1] Statnikov, R. B., et al., *MOVI 1.4 (Multicriteria Optimization and Vector Identification) Software Package,* Certificate of Registration, Registere of Copyright, U.S.A., Registration Number TXU 1-698-418, Date of Registration: May 22, 2010.

[2] Statnikov, R. B., et al., "Multicriteria Analysis Tools in Real-Life Problems," *Journal of Computers & Mathematics with Applications,* Vol. 52, 2006, pp. 1–32.

[3] Statnikov, R. B., A. Bordetsky, and A. Statnikov, "Multicriteria Analysis of Real-Life Engineering Optimization Problems: Statement and Solution," *Nonlinear Analysis,* Vol. 63, 2005, pp. e685–e696.

[4] Statnikov, R., et al., "Definition of the Feasible Solution Set in Multicriteria Optimization Problems with Continuous, Discrete, and Mixed Design Variables," *Nonlinear Analysis,* Vol. 71, No. 12, 2009, pp. e109–e117.

[5] Statnikov, R., A. Bordetsky, and A. Statnikov, "Management of Constraints in Optimization Problems," *Nonlinear Analysis,* Vol. 71, No. 12, 2009, pp. e967–e971.

[6] Statnikov, R., et al., "Visualization Tools for Multicriteria Analysis of the Prototype Improvement Problem," *Proceedings of the First IEEE Symposium on Computational Intelligence in Multi-Criteria Decision-Making (MCDM 2007),* Honolulu, HI, 2007.

[7] Statnikov, R. B., et al., "Visualization Approaches for the Prototype Improvement Problem," *Journal of Multi-Criteria Decision Analysis,* No. 15, 2008, pp. 45–61.

4

Improving Optimal Solutions

In Chapters 2 and 3 we defined criteria constraints and functional constraints and constructed the feasible solution and Pareto optimal solution sets [1, 2]. In this chapter, we will consider a question how optimization results can be improved owing to the correction of the design variable constraints [2–7].

4.1 Solving a New Optimization Problem

4.1.1 New Design Variable Constraints

From the tables of feasible solutions (Figure 3.11) and histograms of the first (Figure 3.12), third (Figure 3.14), and fourth (Figure 3.15) design variables, it follows that we can formulate new boundaries for these design variables (see Table 4.1). The ranges of change for the second and fifth design variables are kept unchanged ($4.0 \cdot 10^4 \le \alpha_2 \le 5.0 \cdot 10^4$; $80 \le \alpha_5 \le 120$). The former rigid functional constraints and criteria constraints are preserved as well. As a result of the correction of design variable constraints, 194 feasible solutions and 25 Pareto optimal solutions were obtained.

Previously in the problem Oscillator, we obtained eight feasible solutions and four Pareto optimal solutions (see Figure 3.7). At the expense of the correction of the specified design variable constraints in problem Oscillator 1, the number of feasible solutions and Pareto optimal solutions has increased to 194 and 25, respectively. Some results of solving this problem are shown in Figures 4.1 through 4.18.

Chapter 3 and this chapter include tables, histograms, and graphs. In Chapter 3, material is presented considering the initial constraints on design

Table 4.1
Intervals of Variation of Design Variables

Initial Intervals (Problem Oscillator)	Subintervals Where the Feasible Solutions Belong (Problem Oscillator)	New Intervals (New Problem Oscillator 1)
$1.1 \cdot 10^6 \le \alpha_1 \le 2.0 \cdot 10^6$	$1.1 \cdot 10^6 \le \alpha_1 \le 1.17 \cdot 10^6$	$9.5 \cdot 10^5 \le \alpha_1 \le 1.25 \cdot 10^6$
$950 \le \alpha_3 \le 1{,}050$	$951 \le \alpha_3 \le 975$	$900 \le \alpha_3 \le 1{,}000$
$30 \le \alpha_4 \le 70$	$42.7 \le \alpha_4 \le 60.35$	$40 \le \alpha_4 \le 65$

variables (problem Oscillator), and in this chapter, material is presented after correcting these constraints (problem Oscillator 1). In both cases the functional constraints and criteria constraints are identical. The reader can compare the respective tables, histograms, and graphs in Chapter 3 and this chapter and see how the correction of the initial constraints improved the results of optimization.

4.1.2 Tables of Feasible and Pareto Optimal Solutions

Tables of feasible solutions as well as histograms are important in the following correction of all constraints, particularly design variable constraints. Next we demonstrate tables of criteria and design variables (Figures 4.1 through 4.4).

4.1.3 Histograms of the Distribution of the Feasible Solutions

In comparison with histograms of the initial problem (Figures 3.12 through 3.16), histograms of the corrected problem (Figures 4.5 through 4.9) show uniform distribution of the feasible solutions.

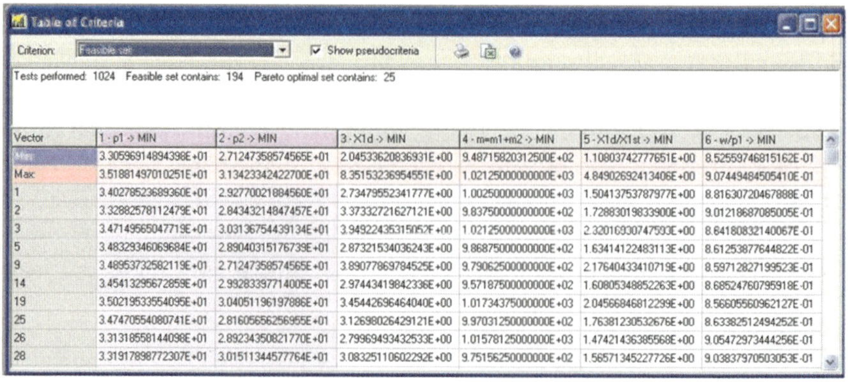
Table of Criteria

Criterion: Feasible set ☑ Show pseudocriteria

Tests performed: 1024 Feasible set contains: 194 Pareto optimal set contains: 25

Vector	1 - p1 -> MIN	2 - p2 -> MIN	3 - X1d -> MIN	4 - m=m1+m2 -> MIN	5 - X1d/X1st -> MIN	6 - w/p1 -> MIN
Min	3.30596914894398E+01	2.71247358574565E+01	2.04533620836931E+00	9.48715820312500E+02	1.10803742777651E+00	8.52559746815162E-01
Max	3.51881497010251E+01	3.13423342422700E+01	8.35153236954551E+00	1.02125000000000E+03	4.84902692413406E+00	9.07449484505410E-01
1	3.40278523689360E+01	2.92770021884560E+01	2.73479552341777E+00	1.00250000000000E+03	1.50413753787977E+00	8.81630720467888E-01
2	3.32882578112479E+01	2.84343214847457E+01	3.37332721627121E+00	9.83750000000000E+02	1.72883019833900E+00	9.01218687085005E-01
3	3.47149565047719E+01	3.03136754439134E+01	3.94922435315052E+00	1.02125000000000E+03	2.32016930747593E+00	8.64180832140067E-01
5	3.48329346069684E+01	2.89040315176739E+01	2.87321534036243E+00	9.86875000000000E+02	1.63414122483113E+00	8.61253877644822E-01
9	3.48953732582119E+01	2.71247358574565E+01	3.89077869784525E+00	9.79062500000000E+02	2.17640433410719E+00	8.59712827199523E-01
14	3.45413295672859E+01	2.99283397714005E+01	2.97443419842336E+00	9.57187500000000E+02	1.60805348852263E+00	8.68524760795918E-01
19	3.50219533554095E+01	3.04051196197886E+01	3.45442696464040E+00	1.01734375000000E+03	2.04566846812299E+00	8.56605560962127E-01
25	3.47470554080741E+01	2.81605656256955E+01	3.12698026429121E+00	9.97031250000000E+02	1.76381230532676E+00	8.63382512494252E-01
26	3.31318558144098E+01	2.89234350821770E+01	2.79969493432533E+00	1.01578125000000E+03	1.47421436385568E+00	9.05472973444256E-01
28	3.31917898772307E+01	3.01511344577764E+01	3.08325110602292E+00	9.75156250000000E+02	1.56571345227726E+00	9.03837970503053E-01

Figure 4.1 Table of criteria: feasible solutions. Fragment.

Table of Criteria

Criterion: Pareto optimal set ☑ Show pseudocriteria

Tests performed: 1024 Feasible set contains: 194 Pareto optimal set contains: 25

Vector	1 - p1 -> MIN	2 - p2 -> MIN	3 - X1d -> MIN	4 - m=m1+m2 -> MIN	5 - X1d/X1st -> MIN	6 - w/p1 -> MIN
Min	3.30732193449298E+01	2.83405264241461E+01	2.04533620836931E+00	9.48715820312500E+02	1.10803742777651E+00	8.52559746815162E-01
Max	3.51881497010251E+01	3.12382655586961E+01	7.96232123851315E+00	1.01497558593750E+03	4.35711592382895E+00	9.07078312731569E-01
117	3.48947303958375E+01	2.96512972424981E+01	2.23274963479056E+00	1.00230468750000E+03	1.28819187913502E+00	8.59728665609023E-01
126	3.48474884360768E+01	2.90896966099589E+01	2.78399441319015E+00	9.56992187500000E+02	1.52793443380162E+00	8.60894180509769E-01
130	3.36573300918093E+01	2.97096519433535E+01	3.31134722371459E+00	9.53964843750000E+02	1.69900569466762E+00	8.91336297863408E-01
242	3.38815139751356E+01	2.92973758876430E+01	2.53846782536874E+00	9.64316406250000E+02	1.32327551287289E+00	8.85438591144890E-01
260	3.30732193449298E+01	3.00145633662976E+01	4.05022997047573E+00	9.49130859375000E+02	2.00098763873406E+00	9.07078312731569E-01
273	3.51144179444971E+01	3.01961114439703E+01	3.17573891780424E+00	9.49912109375000E+02	1.76247307323061E+00	8.54349915394266E-01
313	3.49710101226565E+01	2.90963897171174E+01	2.47350368908885E+00	9.82333984375000E+02	1.40173260818189E+00	8.57853401854241E-01
390	3.43317907791132E+01	2.86573027986895E+01	3.27195027148577E+00	9.53134765625000E+02	1.74109931927011E+00	8.73825667673923E-01
397	3.49747175035373E+01	3.00850833964778E+01	2.93661171263060E+00	9.98447265625000E+02	1.70030965275262E+00	8.57762467901730E-01
419	3.51245562007118E+01	2.97173586884890E+01	2.40707485994157E+00	1.01368164062500E+03	1.42191365505728E+00	8.54103318162125E-01
497	3.47865429754592E+01	2.97058943811299E+01	2.63216952103843E+00	9.70126953125000E+02	1.47159868241651E+00	8.62402453188983E-01
502	3.44319353858611E+01	2.98502983201946E+01	2.26041898222710E+00	9.67001953125000E+02	1.22137678014283E+00	8.71284162908804E-01
546	3.38274825949737E+01	2.97628123292665E+01	2.15156978485474E+00	9.54282226562500E+02	1.10803742777651E+00	8.86852869283794E-01
619	3.51667721042360E+01	3.06792460838268E+01	5.23728846539291E+00	1.01463378906250E+03	3.14518607597009E+00	8.53078011000798E-01
625	3.48763524713552E+01	2.95574845659944E+01	2.36302723543965E+00	9.71665039062500E+02	1.31939533385022E+00	8.60181695452234E-01
734	3.48320643601416E+01	3.12382655586961E+01	7.96232123851315E+00	9.48715820312500E+02	4.35711592382895E+00	8.61275395274278E-01
756	3.34136688728054E+01	2.93073768275213E+01	3.19177910196822E+00	9.52231445312500E+02	1.60446122337709E+00	8.97836155442850E-01
798	3.45505618221438E+01	2.92404076807688E+01	2.68296977385443E+00	9.65170898437500E+02	1.46423599425639E+00	8.68292682313857E-01
806	3.43838707464156E+01	2.94722541987216E+01	2.48986104130836E+00	9.60092773437500E+02	1.32966846722605E+00	8.72502116508432E-01
843	3.51881497010251E+01	3.00264540657824E+01	3.32871507786801E+00	1.01497558593750E+03	1.99219046432755E+00	8.52559746815162E-01
873	3.51476472130789E+01	3.08591810209888E+01	5.75412138005386E+00	9.57553710937500E+02	3.24146964656647E+00	8.53542196384530E-01
878	3.43233494969960E+01	2.92917493423032E+01	2.25291485959694E+00	9.79428710937500E+02	1.22689255024827E+00	8.74040571204324E-01
929	3.43033672558211E+01	2.99965975408494E+01	2.04533620836931E+00	9.95541992187500E+02	1.13182593991452E+00	8.74549713334897E-01
945	3.50956692884183E+01	2.83405264241461E+01	3.09735887639658E+00	9.61948242187500E+02	1.72850168937190E+00	8.54806322496894E-01
986	3.41767598148843E+01	2.98533994423989E+01	2.63129767105792E+00	9.57846679687500E+02	1.39132433983819E+00	8.77789473387549E-01

Figure 4.2 Table of criteria: Pareto optimal solutions.

Design Variable Values

Vectors from: 1 to: 2 Feasible set Show ☐ Hide Constant Variables

Tests performed 1024 Feasible set contains: 194 Pareto optimal set contains: 25

Number of vector	Variable # 1 - k1	Variable # 2 - k2	Variable # 3 - m1	Variable # 4 - m2	Variable # 5 - c
Min	[illegible]	4.00097656250000E+04	9.00097656250000E+02	[illegible]	[illegible]
Max	1.20107421875000E+06	4.99804687500000E+04	9.75000000000000E+02	6.46826171875000E+01	1.19921875000000E+02
1	1.10000000000000E+06	4.50000000000000E+04	9.50000000000000E+02	5.25000000000000E+01	1.00000000000000E+02
2	1.02500000000000E+06	4.75000000000000E+04	9.25000000000000E+02	5.87500000000000E+01	9.00000000000000E+01
3	1.17500000000000E+06	4.25000000000000E+04	9.75000000000000E+02	4.62500000000000E+01	1.10000000000000E+02
5	1.13750000000000E+06	4.12500000000000E+04	9.37500000000000E+02	4.93750000000000E+01	8.50000000000000E+01
9	1.11875000000000E+06	4.43750000000000E+04	9.18750000000000E+02	6.03125000000000E+01	1.07500000000000E+02
14	1.08125000000000E+06	4.56250000000000E+04	9.06250000000000E+02	5.09375000000000E+01	1.12500000000000E+02
19	1.18437500000000E+06	4.78125000000000E+04	9.65625000000000E+02	5.17187500000000E+01	1.08750000000000E+02
25	1.12812500000000E+06	4.96875000000000E+04	9.34375000000000E+02	6.26562500000000E+01	1.11250000000000E+02
26	1.05312500000000E+06	4.71875000000000E+04	9.59375000000000E+02	5.64062500000000E+01	8.12500000000000E+01
28	1.01562500000000E+06	4.84375000000000E+04	9.21875000000000E+02	5.32812500000000E+01	1.16250000000000E+02
30	1.09062500000000E+06	4.09375000000000E+04	9.46875000000000E+02	4.70312500000000E+01	1.06250000000000E+02

Figure 4.3 Table of design variables: feasible solutions. Fragment.

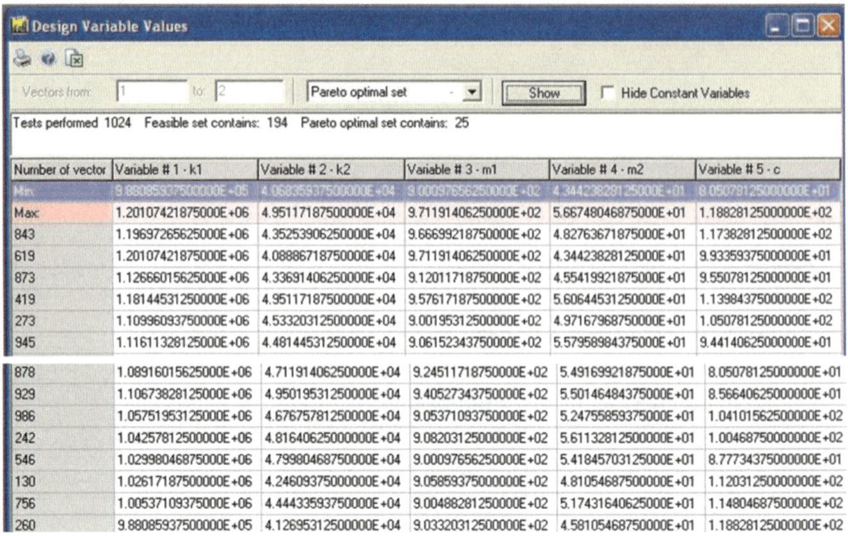

Design Variable Values

Vectors from: 1 to: 2 Pareto optimal set Show Hide Constant Variables

Tests performed 1024 Feasible set contains: 194 Pareto optimal set contains: 25

Number of vector	Variable # 1 - k1	Variable # 2 - k2	Variable # 3 - m1	Variable # 4 - m2	Variable # 5 - c
Min:	9.88085937500000E+05	4.06835937500000E+04	9.00097656250000E+02	4.34423828125000E+01	8.05078125000000E+01
Max:	1.20107421875000E+06	4.95117187500000E+04	9.71191406250000E+02	5.66748046875000E+01	1.18828125000000E+02
843	1.19697265625000E+06	4.35253906250000E+04	9.66699218750000E+02	4.82763671875000E+01	1.17382812500000E+02
619	1.20107421875000E+06	4.08886718750000E+04	9.71191406250000E+02	4.34423828125000E+01	9.93359375000000E+01
873	1.12666015625000E+06	4.33691406250000E+04	9.12011718750000E+02	4.55419921875000E+01	9.55078125000000E+01
419	1.18144531250000E+06	4.95117187500000E+04	9.57617187500000E+02	5.60644531250000E+01	1.13984375000000E+02
273	1.10996093750000E+06	4.53320312500000E+04	9.00195312500000E+02	4.97167968750000E+01	1.05078125000000E+02
945	1.11611328125000E+06	4.48144531250000E+04	9.06152343750000E+02	5.57958984375000E+01	9.44140625000000E+01
878	1.08916015625000E+06	4.71191406250000E+04	9.24511718750000E+02	5.49169921875000E+01	8.05078125000000E+01
929	1.10673828125000E+06	4.95019531250000E+04	9.40527343750000E+02	5.50146484375000E+01	8.56640625000000E+01
986	1.05751953125000E+06	4.67675781250000E+04	9.05371093750000E+02	5.24755859375000E+01	1.04101562500000E+02
242	1.04257812500000E+06	4.81640625000000E+04	9.08203125000000E+02	5.61132812500000E+01	1.00468750000000E+02
546	1.02998046875000E+06	4.79980468750000E+04	9.00097656250000E+02	5.41845703125000E+01	8.77734375000000E+01
130	1.02617187500000E+06	4.24609375000000E+04	9.05859375000000E+02	4.81054687500000E+01	1.12031250000000E+02
756	1.00537109375000E+06	4.44433593750000E+04	9.00488281250000E+02	5.17431640625000E+01	1.14804687500000E+02
260	9.88085937500000E+05	4.12695312500000E+04	9.03320312500000E+02	4.58105468750000E+01	1.18828125000000E+02

Figure 4.4 Table of design variables: Pareto optimal solutions. Fragment.

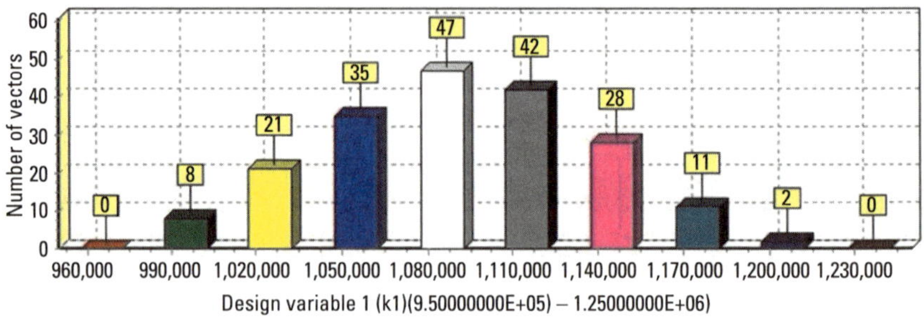

Figure 4.5 First design variable.

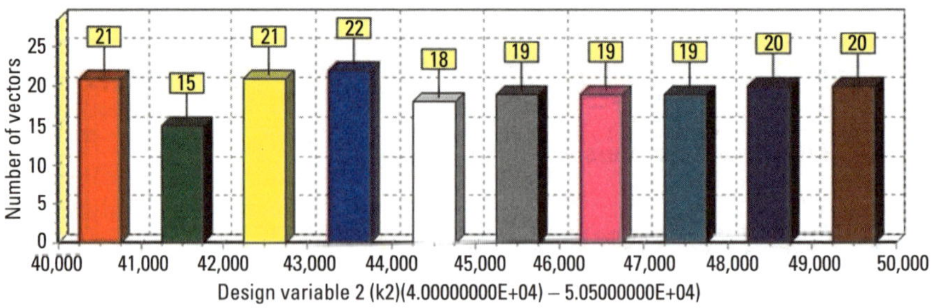

Figure 4.6 Second design variable.

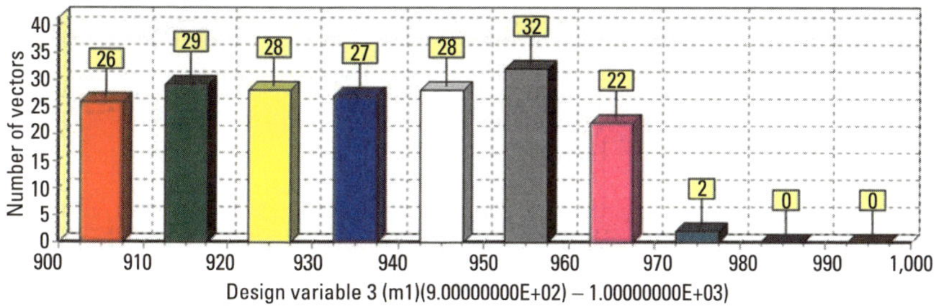

Figure 4.7 Third design variable.

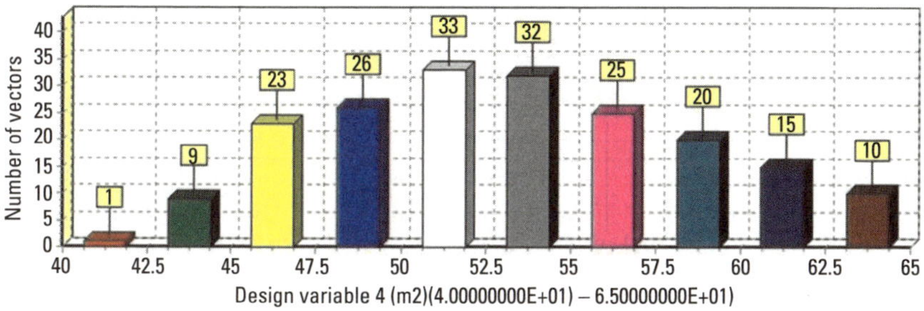

Figure 4.8 Fourth design variable.

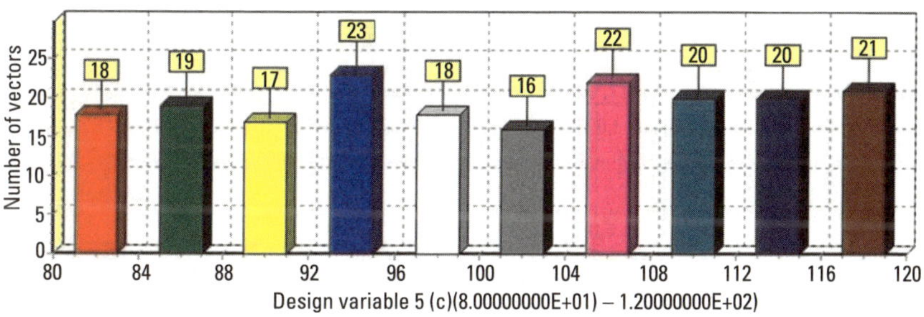

Figure 4.9 Fifth design variable.

4.1.4 Graphs of Criterion Versus Design Variable II

In comparison with the initial graphs (Figures 3.23 through 3.31), only feasible solutions, including the Pareto optimal solutions, are displayed in the graphs Criterion Versus Design Variable II (Figures 4.10–4.13) and Criterion versus Criterion (see Section 4.1.5). One hundred ninety-four feasible solutions and 25 Pareto optimal solutions are shown in these graphs.

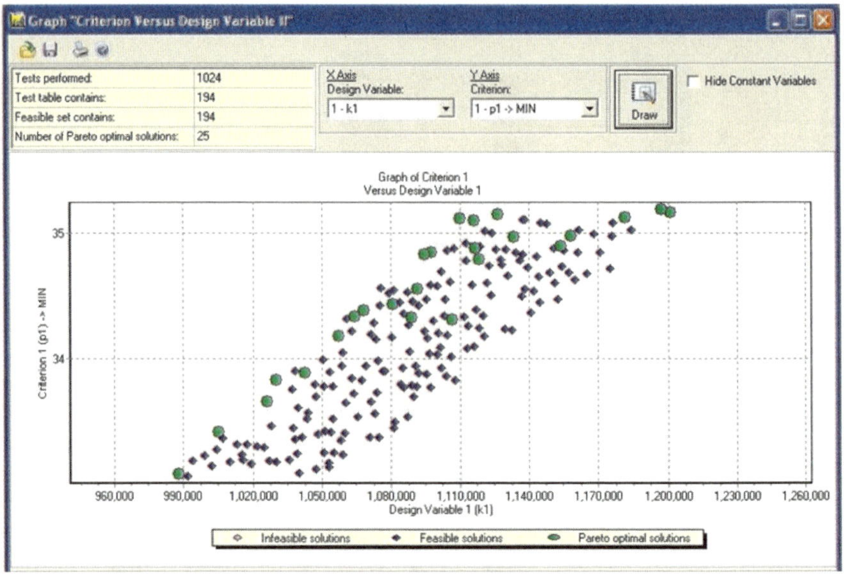

Figure 4.10 First criterion versus first design variable. Green and blue points are feasible solutions; green points are Pareto optimal solutions.

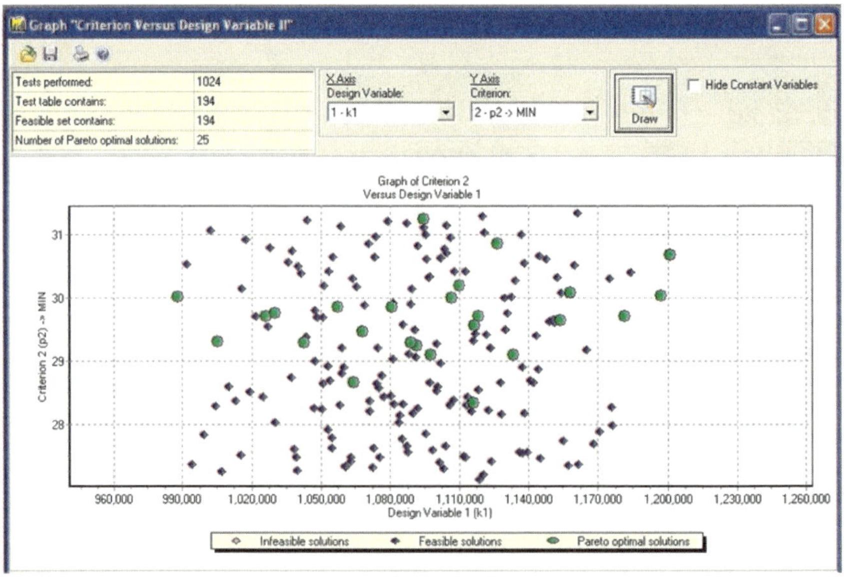

Figure 4.11 Second criterion versus first design variable. Green and blue points are feasible solutions; green points are Pareto optimal solutions.

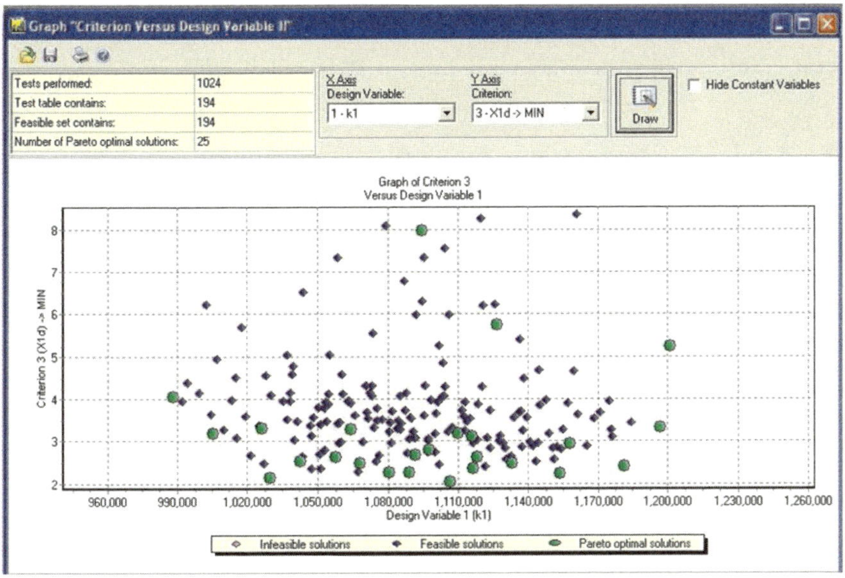

Figure 4.12 Third criterion versus first design variable. Green and blue points are feasible solutions; green points are Pareto optimal solutions.

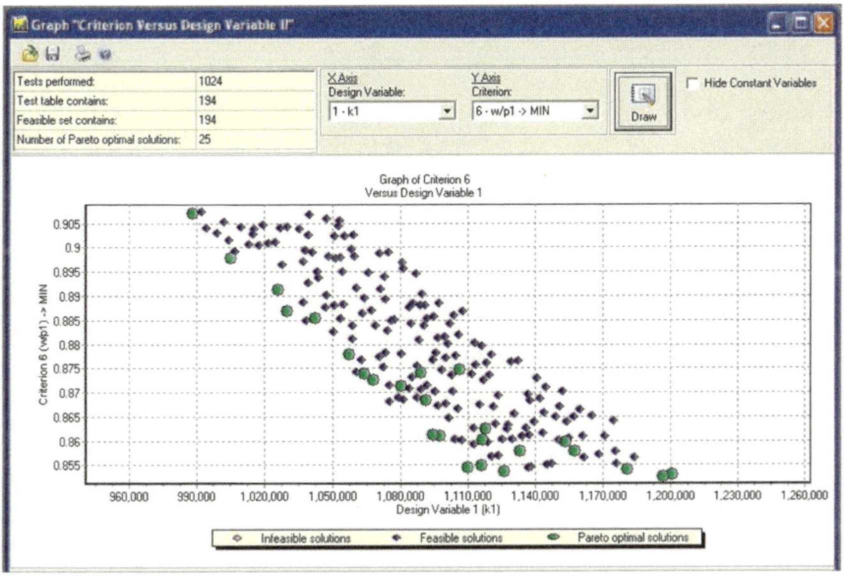

Figure 4.13 Sixth criterion versus first design variable. Green and blue points are feasible solutions; green points are Pareto optimal solutions.

4.1.5 Graphs Criterion Versus Criterion

These graphs are demonstrated in Figures 4.14 through 4.18.

4.2 Construction of the Combined Pareto Optimal Set

4.2.1 Basic Principles

Quite often the analysis of test tables points to advisability of correcting the boundaries of the initial parallelepiped (i.e., redefining the constraints on the design variables) and constructing a new parallelepiped. Suppose that appropriate investigations have been performed in the new parallelepiped, and that the corresponding feasible set has been constructed. Now it is necessary to combine the feasible sets constructed in both these parallelepipeds and define the Pareto optimal solutions in the combined feasible set. The procedure of the repeated correction of parallelepipeds and construction of the combined feasible and Pareto optimal sets is essential for real-life problems. Constructing the combined Pareto optimal set allows us to estimate the contribution of each problem to this set and the expediency of correcting the initial problem.

Assume that the Pareto optimal set P corresponds to the initial statement of the problem (parallelepiped Π) (see Figure 4.19). After correcting the initial statement of problem, we have constructed the new parallelepiped Π_1 and have

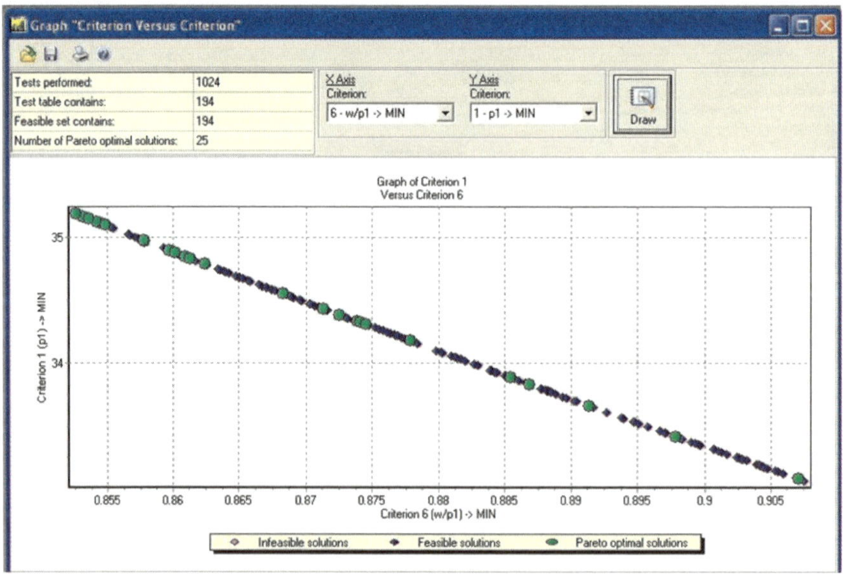

Figure 4.14 First criterion versus sixth criterion. Green and blue points are feasible solutions; green points are Pareto optimal solutions.

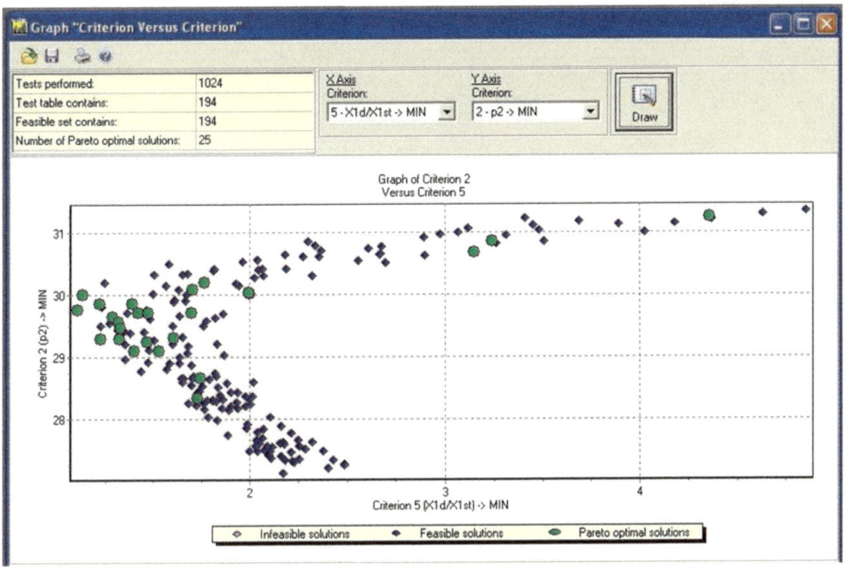

Figure 4.15 Fifth criterion versus second criterion. Green and blue points are feasible solutions; green points are Pareto optimal solutions.

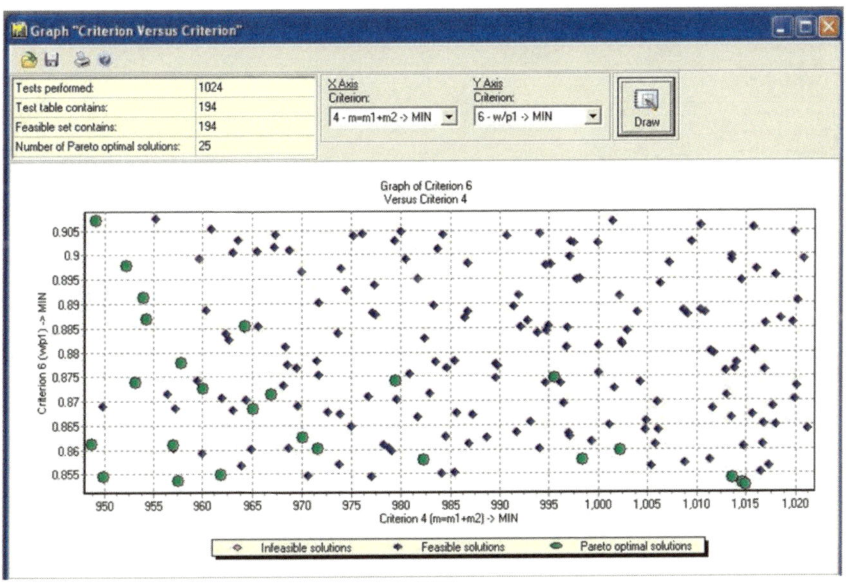

Figure 4.16 Fourth criterion versus sixth criterion. Green and blue points are feasible solutions; green points are Pareto optimal solutions.

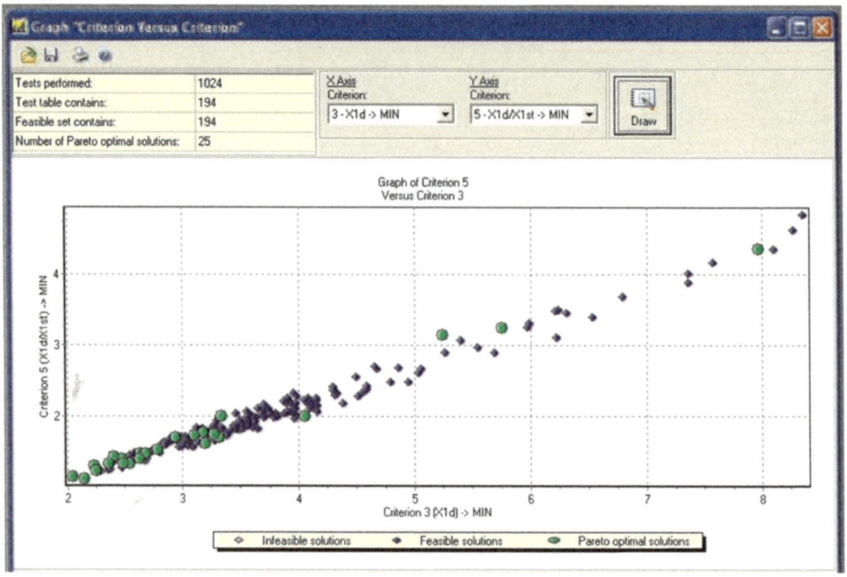

Figure 4.17 Third criterion versus fifth criterion. Green and blue points are feasible solutions; green points are Pareto optimal solutions.

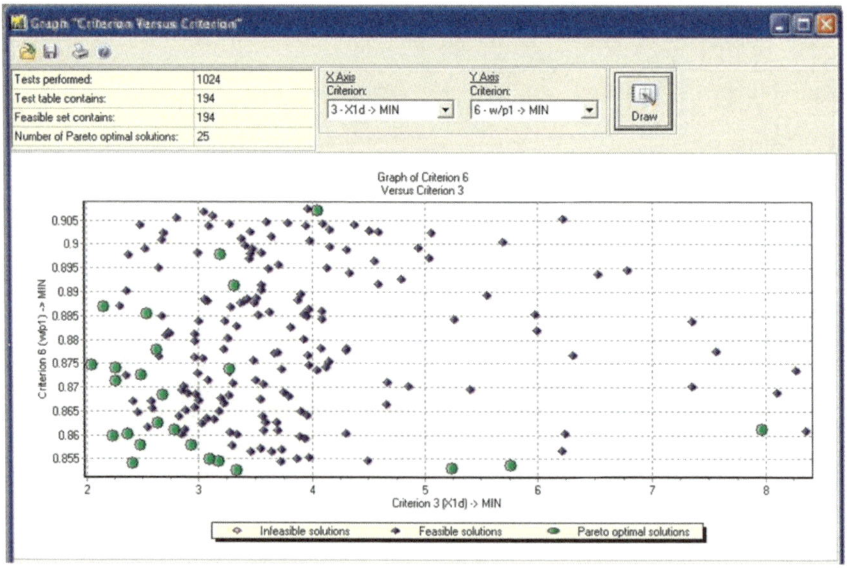

Figure 4.18 Third criterion versus sixth criterion. Green and blue points are feasible solutions; green points are Pareto optimal solutions.

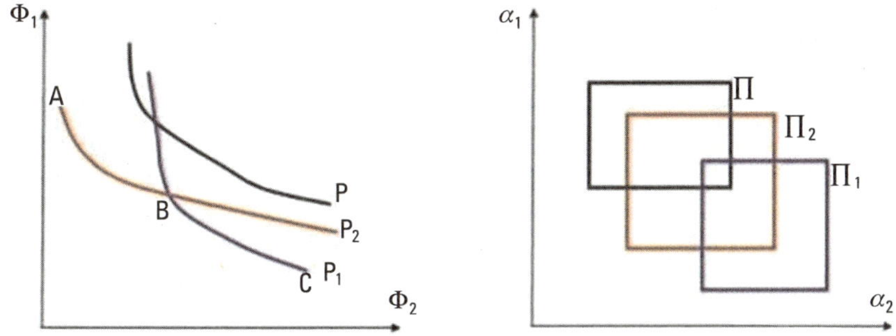

Figure 4.19 Improving the Pareto optimal solutions after correcting the parallelepipeds.

obtained a Pareto optimal set P_1. After analyzing results, we have constructed another new parallelepiped Π_2 and have obtained a Pareto optimal set P_2. Then we have constructed the combined Pareto optimal set presented by two curves AB and BC. One can see that the initial Pareto optimal set P was improved.

There is another situation where construction of the combined feasible and Pareto sets is needed. These are the problems where calculation of the criteria vector for one test requires a significant amount of computer time. Similarly, these are problems that require us to perform a very large number of tests. In these cases, for the construction of the combined Pareto optimal set, we use several computers. For more details, see Chapters 5 and 7.

4.2.2 Tables of Combined Pareto Optimal Solutions

In our cases we have obtained the Pareto optimal solutions in the problem oscillator and problem oscillator 1. The results of constructing the combined Pareto optimal set are shown in Figures 4.20 through 4.22.

4.2.3 Analysis of the Combined Pareto Optimal Set

Generally, to improve the statement and solution of an optimization problem, we use test tables and other tools for multicriteria analysis such as tables of criteria and design variables, tables of functional failures, the histograms, and the graphs of Criterion versus Design Variable II, Criterion versus Criterion, and Criterion versus Design Variable I.

We have obtained the combined Pareto optimal set (Figures 4.20 through 4.22). This set contains 25 Pareto solutions. Then we should perform analysis of the criteria table (see Figure 4.21) and define the best solution. The best solutions by each criterion are marked in Figure 4.21.

In our case the combined Pareto optimal set contains 25 solutions belonging only to the new problem. Thus, all Pareto optimal solutions belonging to

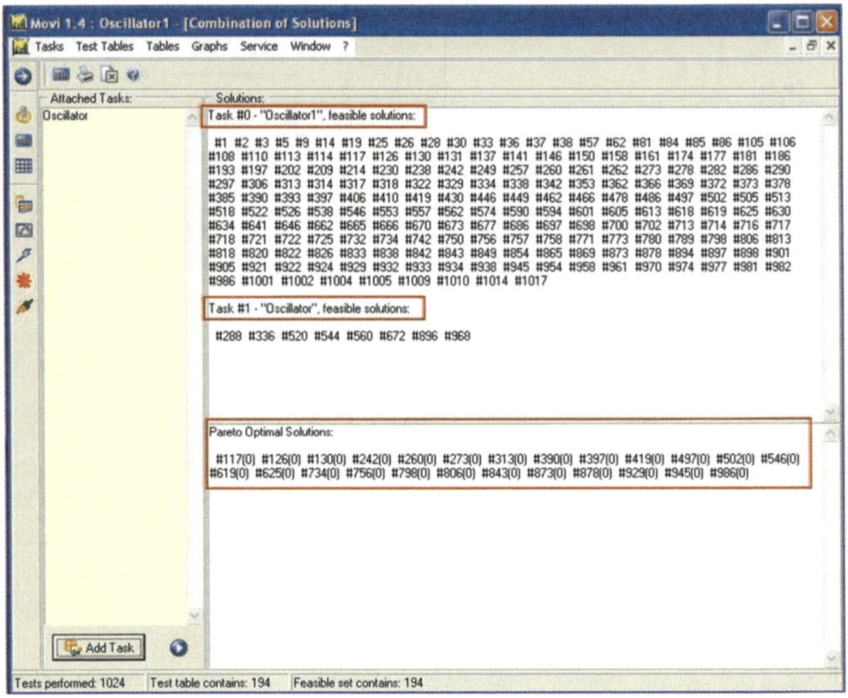

Figure 4.20 The combined Pareto optimal set contains 25 vectors: 117(0), 126(0), 1,130(0), ..., 986(0) belonging to the new problem Oscillator 1. Designations Task #0 and Task #1 correspond to the new problem Oscillator 1 and the initial problem Oscillator, respectively.

the initial problem were improved. As can be seen from Table 4.2, the values of criteria are essentially improved in result of correcting the initial problem. Recall that we seek to minimize all criteria Φ_3, Φ_4, Φ_5, Φ_6. Since Φ_1, Φ_2 are pseudo-criteria, they are not presented in Table 4.2. We have improved former the Pareto optimal solutions on the basis of the correction of the design variable constraints.

4.2.4 Conclusions for Chapters 2 Through 4

Part I gave an overview of the PSI method and MOVI software system. We have also analyzed typical difficulties accompanying the statement of real-life optimization problems. To briefly summarize:

1. We have shown how to correctly state and solve real-life optimization problems. We have considered the determination of criteria constraints and constraints on design variables and functional dependences.

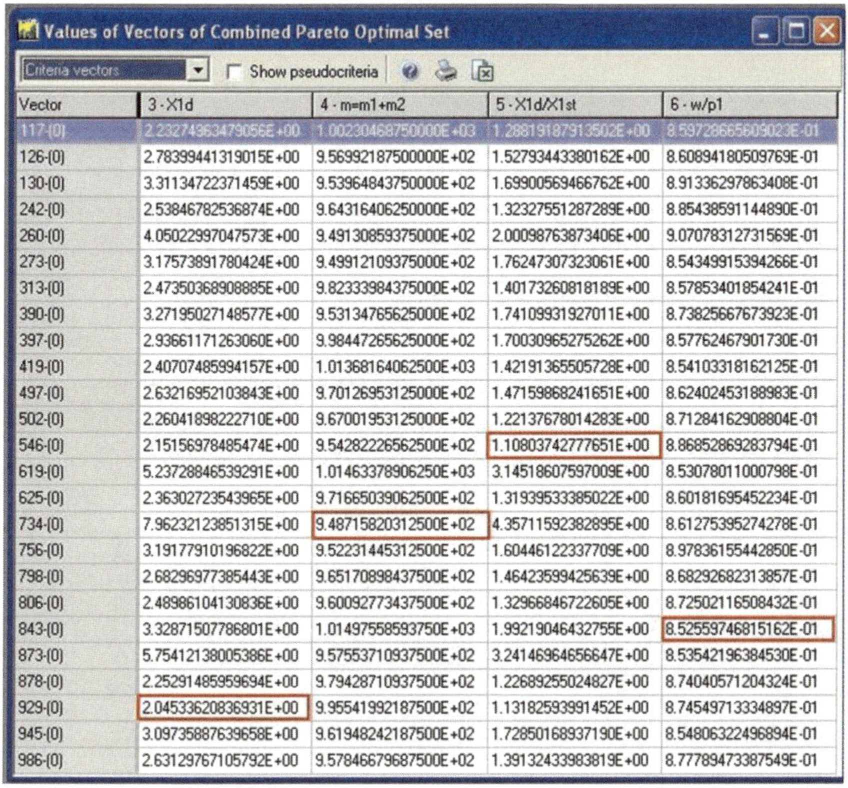

Vector	3 - X1d	4 - m=m1+m2	5 - X1d/X1st	6 - w/p1
117-(0)	2.23274363479056E+00	1.00230468750000E+03	1.28819187913502E+00	8.59728665609023E-01
126-(0)	2.78399441319015E+00	9.56992187500000E+02	1.52793443380162E+00	8.60894180509769E-01
130-(0)	3.31134722371459E+00	9.53964843750000E+02	1.69900569466762E+00	8.91336297863408E-01
242-(0)	2.53846782536874E+00	9.64316406250000E+02	1.32327551287289E+00	8.85438591144890E-01
260-(0)	4.05022997047573E+00	9.49130859375000E+02	2.00098763873406E+00	9.07078312731569E-01
273-(0)	3.17573891780424E+00	9.49912109375000E+02	1.76247307323061E+00	8.54349915394266E-01
313-(0)	2.47350368908885E+00	9.82333984375000E+02	1.40173260818189E+00	8.57853401854241E-01
390-(0)	3.27195027148577E+00	9.53134765625000E+02	1.74109931927011E+00	8.73825667673923E-01
397-(0)	2.93661171263060E+00	9.98447265625000E+02	1.70030965275262E+00	8.57762467901730E-01
419-(0)	2.40707485994157E+00	1.01368164062500E+03	1.42191365505728E+00	8.54103318162125E-01
497-(0)	2.63216952103843E+00	9.70126953125000E+02	1.47159868241651E+00	8.62402453188983E-01
502-(0)	2.26041898222710E+00	9.67001953125000E+02	1.22137678014283E+00	8.71284162908804E-01
546-(0)	2.15156978485474E+00	9.54282226562500E+02	1.10803742777651E+00	8.86852869283794E-01
619-(0)	5.23728846539291E+00	1.01463378906250E+03	3.14518607597009E+00	8.53078011000798E-01
625-(0)	2.36302723543965E+00	9.71665039062500E+02	1.31939533385022E+00	8.60181695452234E-01
734-(0)	7.96232123851315E+00	9.48715820312500E+02	4.35711592382895E+00	8.61275395274278E-01
756-(0)	3.19177910196822E+00	9.52231445312500E+02	1.60446122337709E+00	8.97836155442850E-01
798-(0)	2.68296977385443E+00	9.65170898437500E+02	1.46423599425639E+00	8.68292682313857E-01
806-(0)	2.48986104130836E+00	9.60092773437500E+02	1.32966846722605E+00	8.72502116508432E-01
843-(0)	3.32871507786801E+00	1.01497558593750E+03	1.99219046432755E+00	8.52559746815162E-01
873-(0)	5.75412138005386E+00	9.57553710937500E+02	3.24146964656647E+00	8.53542196384530E-01
878-(0)	2.25291485959694E+00	9.79428710937500E+02	1.22689255024827E+00	8.74040571204324E-01
929-(0)	2.04533620836931E+00	9.95541992187500E+02	1.13182593991452E+00	8.74549713334897E-01
945-(0)	3.09735887639658E+00	9.61948242187500E+02	1.72850168937190E+00	8.54806322496894E-01
986-(0)	2.63129767105792E+00	9.57846679687500E+02	1.39132433983819E+00	8.77789473387549E-01

Figure 4.21 Criteria table.

2. We have demonstrated that the formulation and solution of the real-life problem constitute a single process; this process should be interactive.
3. Multicriteria analysis tools such as test tables, tables of feasible and Pareto optimal solutions, design variable histograms and criteria histograms, graphs Criterion versus Design Variable, graphs Criterion versus Criterion, and tables of functional failures for correcting functional constraints have been described.
4. For a definition of the "soft" constraints on functional dependences, these dependences have been presented in the form of pseudo-criteria.
5. We have shown that the analysis of test tables and tables of functional failures allows one to understand how the functional constraints work and, if necessary, to correct them.

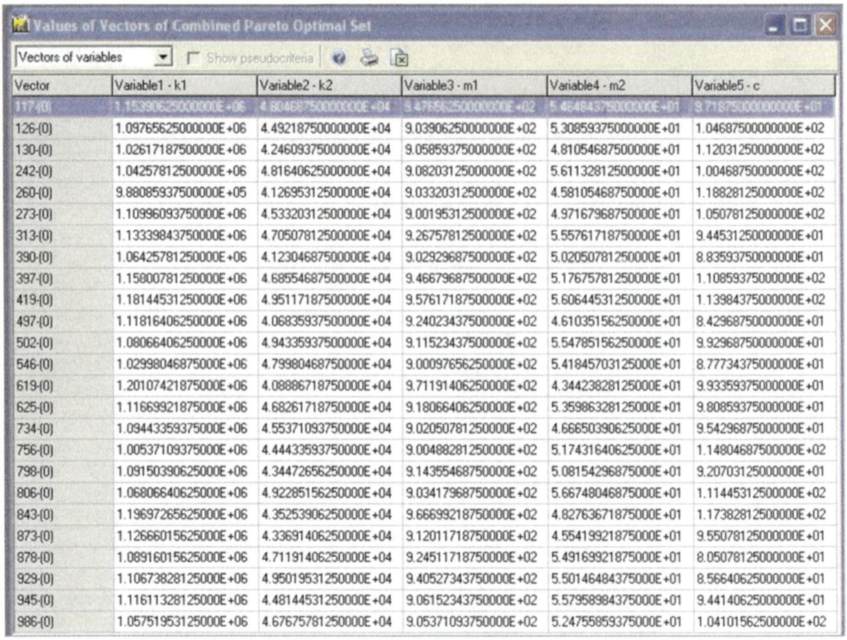

Vector	Variable1 - k1	Variable2 - k2	Variable3 - m1	Variable4 - m2	Variable5 - c
117-(0)	1.15390625000000E+06	4.80468750000000E+04	9.47656250000000E+02	5.46484375000000E+01	9.71875000000000E+01
126-(0)	1.09765625000000E+06	4.49218750000000E+04	9.03906250000000E+02	5.30859375000000E+01	1.04687500000000E+02
130-(0)	1.02617187500000E+06	4.24609375000000E+04	9.05859375000000E+02	4.81054687500000E+01	1.12031250000000E+02
242-(0)	1.04257812500000E+06	4.81640625000000E+04	9.08203125000000E+02	5.61132812500000E+01	1.00468750000000E+02
260-(0)	9.88085937500000E+05	4.12695312500000E+04	9.03320312500000E+02	4.58105468750000E+01	1.18828125000000E+02
273-(0)	1.10996093750000E+06	4.53320312500000E+04	9.00195312500000E+02	4.97167968750000E+01	1.05078125000000E+02
313-(0)	1.13339843750000E+06	4.70507812500000E+04	9.26757812500000E+02	5.55761718750000E+01	9.44531250000000E+01
390-(0)	1.06425781250000E+06	4.12304687500000E+04	9.02929687500000E+02	5.02050781250000E+01	8.83593750000000E+01
397-(0)	1.15800781250000E+06	4.68554687500000E+04	9.46679687500000E+02	5.17675781250000E+01	1.10859375000000E+02
419-(0)	1.18144531250000E+06	4.95117187500000E+04	9.57617187500000E+02	5.60644531250000E+01	1.13984375000000E+02
497-(0)	1.11816406250000E+06	4.06835937500000E+04	9.24023437500000E+02	4.61035156250000E+01	8.42968750000000E+01
502-(0)	1.08066406250000E+06	4.94335937500000E+04	9.11523437500000E+02	5.54785156250000E+01	9.92968750000000E+01
546-(0)	1.02998046875000E+06	4.79980468750000E+04	9.00097656250000E+02	5.41845703125000E+01	8.77734375000000E+01
619-(0)	1.20107421875000E+06	4.08886718750000E+04	9.71191406250000E+02	4.34423828125000E+01	9.93359375000000E+01
625-(0)	1.11669921875000E+06	4.68261718750000E+04	9.18066406250000E+02	5.35986328125000E+01	9.80859375000000E+01
734-(0)	1.09443359375000E+06	4.55371093750000E+04	9.02050781250000E+02	4.66650390625000E+01	9.54296875000000E+01
756-(0)	1.00537109375000E+06	4.44433593750000E+04	9.00488281250000E+02	5.17431640625000E+01	1.14804687500000E+02
798-(0)	1.09150390625000E+06	4.34472656250000E+04	9.14355468750000E+02	5.08154296875000E+01	9.20703125000000E+01
806-(0)	1.06806640625000E+06	4.92285156250000E+04	9.03417968750000E+02	5.66748046875000E+01	1.11445312500000E+02
843-(0)	1.19697265625000E+06	4.35253906250000E+04	9.66699218750000E+02	4.82763671875000E+01	1.17382812500000E+02
873-(0)	1.12666015625000E+06	4.33691406250000E+04	9.12011718750000E+02	4.55419921875000E+01	9.55078125000000E+01
878-(0)	1.08916015625000E+06	4.71191406250000E+04	9.24511718750000E+02	5.49169921875000E+01	8.05078125000000E+01
929-(0)	1.10673828125000E+06	4.95019531250000E+04	9.40527343750000E+02	5.50146484375000E+01	8.56640625000000E+01
945-(0)	1.11611328125000E+06	4.48144531250000E+04	9.06152343750000E+02	5.57958984375000E+01	9.44140625000000E+01
986-(0)	1.05751953125000E+06	4.67675781250000E+04	9.05371093750000E+02	5.24755859375000E+01	1.04101562500000E+02

Figure 4.22 Table of designs.

Table 4.2
Comparison of Results

Problems	Best Value of Criterion 3	Best Value of Criterion 4	Best Value of Criterion 5	Best Value of Criterion 6
Pareto optimal set from initial problem (Figure 3.8)	2.932 (Solution 288)	1,002.77 (Solution 336)	1.636 (Solution 288)	0.86 (Solution 520)
Combined Pareto optimal set (contains solutions only from new problem) (Figure 4.21)	2.045 (Solution 929)	9.487 (Solution 734)	1.108 (Solution 546)	0.852 (Solution 843)

6. We have presented the process of improving the Pareto optimal solutions via an analysis of the feasible solution set and following the correction of the design variable constraints.
7. The construction of the combined Pareto optimal set has been described.

8. Examples have demonstrated the sensitivity of the feasible solution and Pareto optimal sets to the change of constraints.
9. Finally, we have discussed that after the construction of feasible solution set, an expert can consider employing single-criterion optimization methods for the further improvement of the obtained results (e.g., the value of any criterion [6]).

In Part II, we focus on the formulation and solution of real-life problems such as multicriteria design, multicriteria identification, mutlicriteria design of controlled engineering systems, the search for the compromise solution when the desired solution is unattainable, the multicriteria analysis from observational data, the multicriteria optimization of large-scale systems, adopting PSI methods for a database search, and the multicriteria analysis of an adaptive control system. For each of these problems, we describe the value proposition of using multicriteria analysis with the PSI method and provide empirical results.

References

[1] Statnikov, R. B., and J. B. Matusov, *Multicriteria Analysis in Engineering,* Dordrecht/Boston/London: Kluwer Academic Publishers, 2002.

[2] Statnikov, R. B., et al., *MOVI 1.4 (Multicriteria Optimization and Vector Identification) Software Package,* Certificate of Registration, Registere of Copyright, U.S.A., Registration Number TXU 1-698-418, Date of Registration: May 22, 2010.

[3] Statnikov, R. B., et al., "Multicriteria Analysis Tools in Real-Life Problems," *Journal of Computers & Mathematics with Applications,* Vol. 52, 2006, pp. 1–32.

[4] Statnikov, R. B., A. Bordetsky, and A. Statnikov, "Multicriteria Analysis of Real-Life Engineering Optimization Problems: Statement and Solution," *Nonlinear Analysis,* Vol. 63, 2005, pp. e685–e696.

[5] Statnikov, R., A. Bordetsky, and A. Statnikov, "Management of Constraints in Optimization Problems," *Nonlinear Analysis,* Vol. 71, No. 12, 2009, pp. e967–e971.

[6] Statnikov, R., et al., "Definition of the Feasible Solution Set in Multicriteria Optimization Problems with Continuous, Discrete, and Mixed Design Variables," *Nonlinear Analysis,* Vol. 71, No. 12, 2009, pp. e109–e117.

[7] Statnikov, R. B., et al., "Visualization Approaches for the Prototype Improvement Problem," *Journal of Multi-Criteria Decision Analysis,* No. 15, 2008, pp. 45–61.

Part II
Applications to Real-Life Problems

In this part we will show the statement and solution of the main engineering optimization problems such as design, identification, operational development, and multicriteria analysis from observational data.

A crucial question in engineering optimization is the adequacy of the mathematical model to the actual object. Without estimating the model's adequacy, the search for optimal design variables has no applied sense, but what is the measure of adequacy? To what extent can we trust one model or another? In this connection we will discuss identification problems.

One of the basic engineering optimization problems is the improvement of the prototype. This problem is often encountered by industrial and academic organizations that produce and design various objects (e.g., motor vehicles, machine tools, ships, bridges, and aircraft).

For many applied optimization problems, it is necessary to carry out a large-scale numerical experiment in order to construct the feasible set. For this reason, a search for optimal solutions is often not carried out at all. We will mention the main types of computationally expensive problems and show how to solve them.

5

Multicriteria Design

Applying the PSI method in multicriteria design has been described in numerous publications [1–22]. In this chapter we demonstrate examples of multicriteria design: *improving the* ship prototype, truck frame prototype, rear axle housing, and orthotropic bridge.

5.1 Multicriteria Analysis of the Ship Design Prototype

5.1.1 Improvement Problem

The mathematical model described next is based on the references [2, 14–16, 19–21]. The problem has eight criteria that are defined implicitly. Two of the eight are performance criteria: resistance performance defined using Fung's resistance prediction algorithm (bare hull residuary resistance coefficient) [20] and the seakeeping performance defined by Bales formula (Bales seakeeping rank) [19]. The other six criteria are pseudo-criteria.

5.1.1.1 Initial (or the First) Statement of the Optimization Problem: Parallelepiped Π_1

The model is based on 14 design variables:

- α_1: Length of the design water line (assumed to be equal to the length between perpendiculars), m;
- α_2: Beam of the design water line (assumed to be equal to the beam amid ships), m;
- α_3: Draft (assumed to be equal to the draft amid ships), m;

- α_4: Distance from station 0 (FP) to the cut-up point, m;
- α_5: Water plane area forward of amid ships, m^2;
- α_6: Water plane area aft of amid ships, m^2;
- α_7: Displaced volume forward of amid ships, m^3;
- α_8: Displaced volume aft of amid ships, m^3;
- α_9: Prismatic coefficient;
- α_{10}: Projected transom area, m^2;
- α_{11}: Projected transom width, m;
- α_{12}: Projected transom depth, m;
- α_{13}: Half entrance angle, degrees;
- α_{14}: Longitudinal center of buoyancy from the fore perpendicular, m.

Thus, the design variable vector is $\alpha = (\alpha_1, \ldots, \alpha_{14})$. Values of the prototype design variables α^p and initial design variable constraints (i.e., parallelepiped Π_1) are provided in Table 5.1.

We want to maximize the performance criterion Φ_1 (seakeeping rank) and to minimize the performance criterion Φ_2 (residuary resistance coefficient). The vector of performance criteria of prototype is $\Phi^p = (8.608;\ 1.968)$ (see Table 5.2).

There are 10 functional relations:

- f1: Displacement, metric ton;
- f2: Block coefficient;
- f3: Maximum section area coefficient (assumed to be equal to the midship section coefficient, cm);
- f4: Water plane area coefficient;
- f5: Water plane area coefficient forward of amidships;
- f6: Water plane area coefficient aft of amidships;
- f7: Draft-to-length ratio;
- f8: Cut-up ratio;
- f9: Vertical prismatic coefficient forward of amidships;
- f10: Vertical prismatic coefficient aft of amidships.

The following rigid constraints are imposed on the above four functional relations:

Table 5.1

Table of Design Variables: Parallelepiped Π_1

Design Variables	Prototype α^p	Π_1 Lower bound	Upper bound	Pareto Optimal Solutions in Initial Statement α^{26087}	α^{75527}	α^{81087}
α_1	90.700	85.700	95.700	94.749	94.485	95.592
α_2	12.670	10.670	14.670	11.659	11.860	11.656
α_3	3.700	3.500	3.900	3.581	3.578	3.679
α_4	51.650	48.650	54.650	52.081	50.611	52.584
α_5	380.10	330.10	430.100	416.938	427.530	404.837
α_6	552.900	502.90	602.900	506.413	505.068	514.845
α_7	991.200	891.20	1,091.200	930.403	952.688	940.701
α_8	1,040.100	940.00	1,140.000	1,041.730	1,066.618	1,096.428
α_9	0.626	0.620	0.635	0.626	0.624	0.625
α_{10}	11.740	9.740	13.740	11.758	10.099	10.416
α_{11}	12.120	10.120	14.120	10.555	10.982	10.622
α_{12}	0.950	0.750	1.050	0.880	1.025	0.754
α_{13}	13.000	12.000	14.000	12.011	13.986	12.607
α_{14}	45.900	43.900	47.900	47.810	47.604	44.024

- f1 ≤ 2,100;
- f2 ≤ 0.51;
- f3 ≥ 0.77;
- f4 ≤ 0.84;
- f4 ≥ 0.80.

The other functional constraints on f5, …, f10 are "soft" (i.e., they can be changed in some limits); however, it is difficult to formulate these constraints a priori. According to the PSI method, the functional relations with "soft" constraints should be interpreted as the pseudo-criteria (i.e., $\Phi_3 = f5$, $\Phi_4 = f6$, $\Phi_5 = f7$, $\Phi_6 = f8$, $\Phi_7 = f9$, and $\Phi_8 = f10$). In this case, the constraints are defined during solution of the problem (on the basis of the analysis of the test table). Values Φ_3, Φ_4, Φ_5,Φ_6 Φ_7, Φ_8 for the prototype are shown in Table 5.2.

Table 5.2
Table of Criteria

Criteria	Prototype Φ^p	Pareto Optimal Solutions in Initial Statement #26087	#75527	#81087	Pareto Optimal Solutions in Second Statement #74223	#49109	#106467
Φ_1 (max)	8.608	14.342	14.267	12.614	14.537	15.4598	15.4591
Φ_2 (min)	1.968	1.847	1.803	1.711	1.694	1.873	1.852
Φ_3 (pseudo)	0.662	0.755	0.763	0.727	0.753	0.767	0.767
Φ_4 (pseudo)	0.962	0.917	0.901	0.924	0.905	0.911	0.911
Φ_5 (pseudo)	0.041	0.038	0.038	0.038	0.038	0.039	0.038
Φ_6 (pseudo)	0.569	0.550	0.536	0.550	0.554	0.565	0.554
Φ_7 (pseudo)	0.705	0.623	0.623	0.632	0.605	0.596	0.603
Φ_8 (pseudo)	0.508	0.574	0.590	0.579	0.578	0.560	0.560
	Pareto Optimal Solutions in Final Statement						
Criteria	**#113487**	**#4145**	**#68410**	**#39801**	**#53988**	**#72461**	**#7511**
Φ_1 (max)	15.308	15.183	15.131	15.052	14.887	14.805	14.601
Φ_2 (min)	1.677	1.674	1.657	1.645	1.644	1.629	1.622
Φ_3 (pseudo)	0.760	0.755	0.760	0.753	0.758	0.752	0.752
Φ_4 (pseudo)	0.918	0.923	0.916	0.926	0.921	0.926	0.928
Φ_5 (pseudo)	0.038	0.038	0.038	0.038	0.038	0.038	0.038
Φ_6 (pseudo)	0.538	0.547	0.558	0.534	0.551	0.538	0.554
Φ_7 (pseudo)	0.598	0.601	0.602	0.597	0.607	0.606	0.615
Φ_8 (pseudo)	0.561	0.553	0.561	0.557	0.559	0.563	0.568

Therefore, to determine the feasible solution set we solve the problem with criteria vector $\Phi = (\Phi_1, \Phi_2, \Phi_3, \Phi_4, \Phi_5, \Phi_6\ \Phi_7, \Phi_8)$. The vector of functional relations is (f1, f2, f3, f4).

The analysis process summarized in the next section is based on the application of the PSI method with LP_τ sequences.

We should construct the Pareto set taking into account only the performance criteria Φ_1 and Φ_2 because Φ_3, Φ_4, Φ_5, Φ_6, Φ_7, Φ_8 are pseudo-criteria.

Since we are solving the problem of improving the prototype, the criteria constraints equal to the values of the prototype criteria ($\mathbf{\Phi}_1^p = \mathbf{\Phi}_1^{**} = 8.608$ and $\mathbf{\Phi}_2^p = \mathbf{\Phi}_2^{**} = 1.968$) were accepted.

In summary, we want to investigate the problem with a 14-dimensional design variable space and an eight-dimensional criteria space, keeping in mind complex constraints, which we need to correct in order to construct feasible solution set.

Next we will show how to solve this problem.

5.1.1.2 Solution of the Initial Statement of the Optimization Problem

We performed N = 131,072 tests in parallelepiped Π_1 and only N1 = 1,487 vectors entered the test table. The 129,585 solutions did not satisfy the functional constraints. Since we are solving the problem of improving the prototype, the performance criteria constraints equal to the values of the prototype ($\Phi_1^p = \Phi_1^{**}$ = 8.608 and $\Phi_2^p = \Phi_2^{**}$ = 1.968) were accepted. Based on the analysis of the test tables, an expert weakened the constraints on the pseudo-criteria Φ_4, Φ_6, and Φ_8 in comparison with the prototype. As a result, NF = 240 vectors (including the prototype) entered the feasible set, and NP = 3 vectors are the Pareto optimal solution set (the remaining 1,247 vectors did not satisfy the criteria constraints). The coefficient of the efficiency of searching the feasible solution set is γ_F = 240/131,072 = 0.0018 (for the Pareto optimal set it is γ_p = 0.00002). Very low values of the coefficient γ_F point out the difficulties in searching the feasible solutions.

These vectors in the criteria space and design variable space are denoted as Φ^{26087}, Φ^{81087}, Φ^{75527} and α^{26087}, α^{81087}, α^{75527}, respectively. To simplify the notation, we will denote the vectors of criteria corresponding to the *i*th test (i.e., Φ^i) simply as #*i*. Therefore, Φ^{26087}, Φ^{81087}, Φ^{75527} are written as #26087, #81087, and #75527 (see Table 5.2). The values of the design variables and the criteria of Pareto optimal solutions are given in Tables 5.1 and 5.2.

Next we briefly show some multicriteria tools that led to an improvement of the present results.

Graphs: Criterion Versus Criterion

Figure 5.1[1] shows dependency between the first criterion (seakeeping rank) and the second criterion (residuary resistance coefficient)[2].

Examples of the dependences between criterion Φ_1 versus pseudo-criterion Φ_3 and criterion Φ_1 versus pseudo-criterion Φ_4 are shown in Figures 5.2 and 5.3.

Graphs: Criterion Versus Design Variable

Figures 5.4 and 5.5 show the dependency of the criterion Φ_1 versus the design variable α_5 and Φ_2 versus the design variable α_{10}. Here we see a linear dependency between the criterion Φ_1 and the design variable α_5 and a more complex relationship between the criterion Φ_2 and the design variable α_{10}.

1. Figures 5.1 through 5.17, 5.19, 7.1 through 7.5, 7.8 through 7.13, 7.15, and 9.3 through 9.13 were produced by MOVI software.
2. On the graphs criterion versus criterion and criterion versus design variable, the feasible solutions (blue and green points), the Pareto optimal solutions (green points), the prototype (red or yellow triangle), and the unfeasible solutions (magenta) are shown (see Figures 5.1 through 5.5, 5.7, 5.8, 5.10 through 5.17, 9.4 through 9.9, and 9.11 through 9.13.

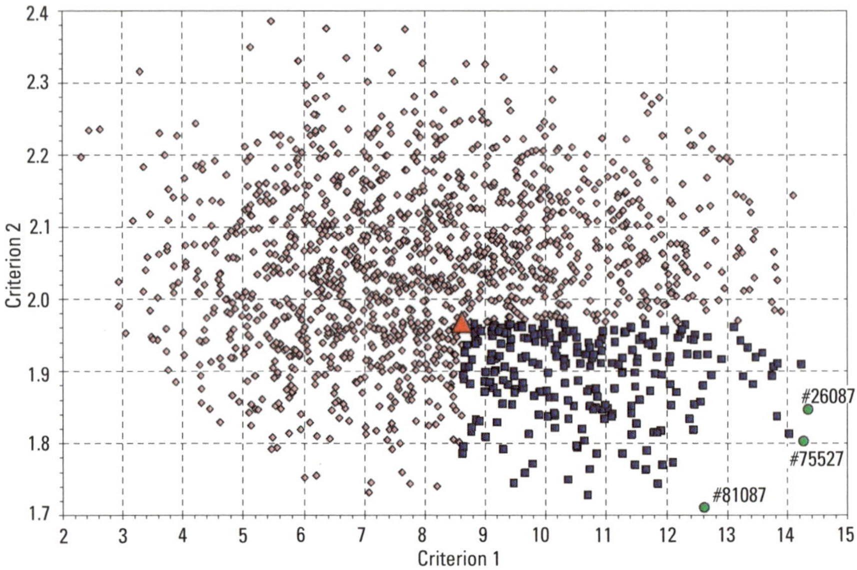

Figure 5.1 The dependency between criteria Φ_1 and Φ_2. Initial statement: Feasible solutions ND = 238; Pareto optimal solutions NP = 3.

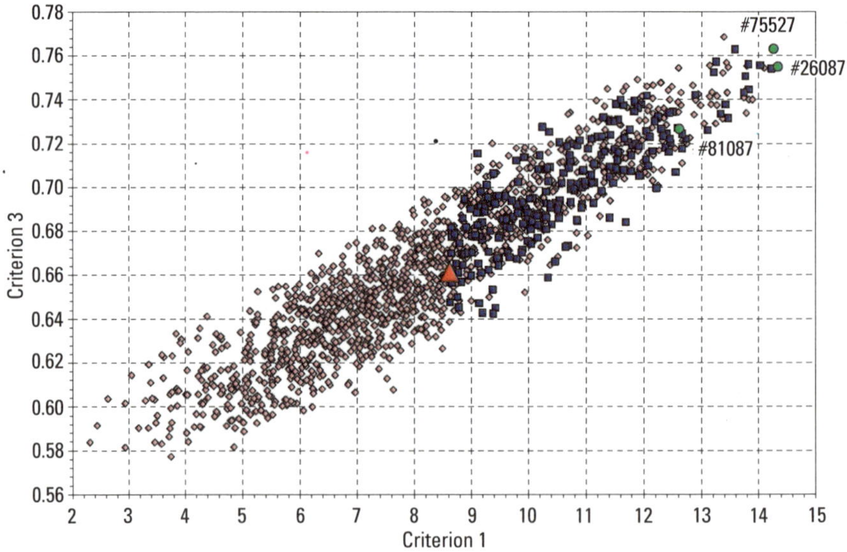

Figure 5.2 The dependency between criterion Φ_1 and pseudo-criterion Φ_3.

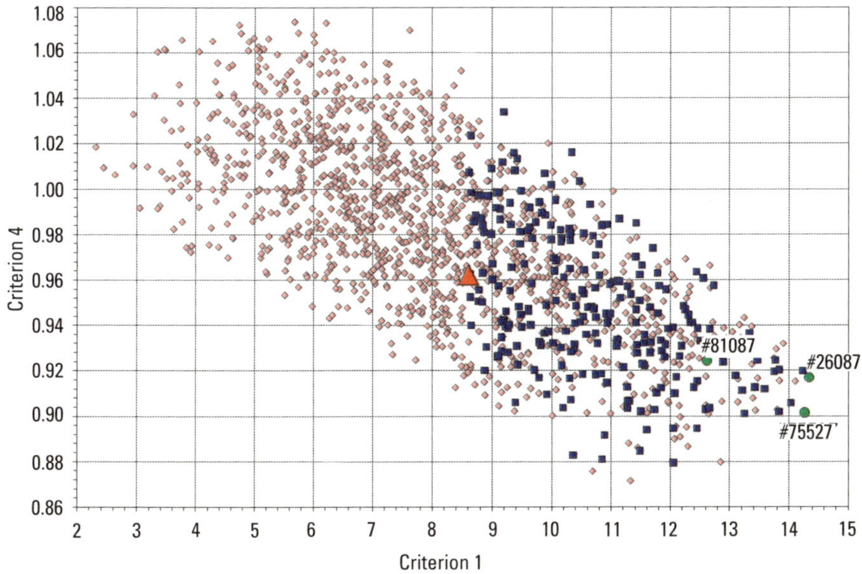

Figure 5.3 The dependency between criterion Φ_1 and pseudo-criterion Φ_4.

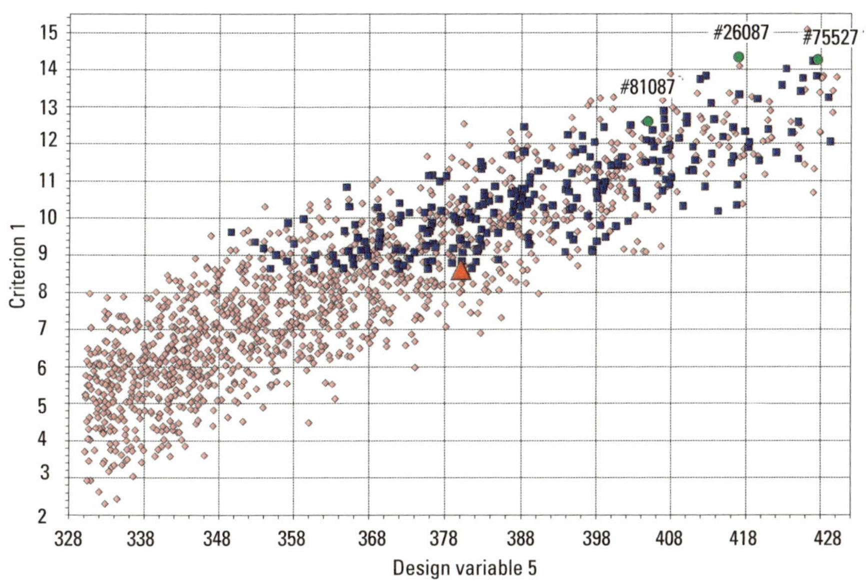

Figure 5.4 The dependency of criterion Φ_1 on design variable α_5.

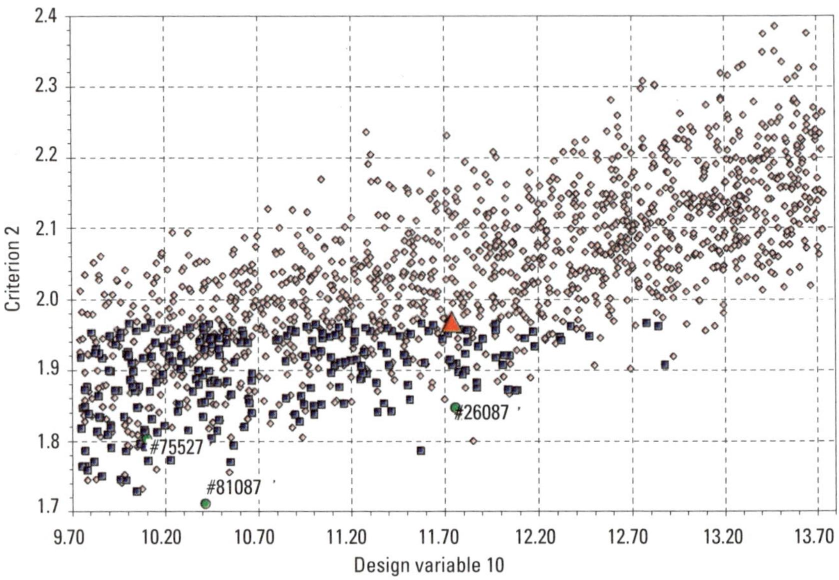

Figure 5.5 The dependency of criterion Φ_2 on design variable α_{10}.

Histograms of Feasible Solutions

The distributions of feasible solutions for the range of the second and tenth design variables in the initial parallelepiped Π_1 are shown in Figure 5.6(a). We can see that there are large "gaps" that do not contain any feasible solutions. Similar distributions with "gaps" are observed for the fifth, sixth, and ninth design variables.

5.1.1.3 Second Statement and Solution of the Optimization Problem: Parallelepiped Π_2

The analysis of the feasible solutions has pointed ways for the correction of the initial problem statement. The reasoning for construction of the parallelepiped Π_2 is given next:

- From the histogram of the second design variable [Figure 5.6(a)], it follows that the value of the upper boundary can be decreased to $\alpha_2^{**} = 13.5$ (recall that in the initial parallelepiped Π_1 we had $\alpha_2^{**} = 14.67$).
- From the graph criterion versus design variable (Figure 5.5) and the histogram of the tenth design variable [Figure 5.6(a)], it follows that the value of the upper boundary can be decreased to $\alpha_{10}^{**} = 13.0$ (recall that in the initial parallelepiped Π_1 we had $\alpha_{10}^{**} = 13.74$).

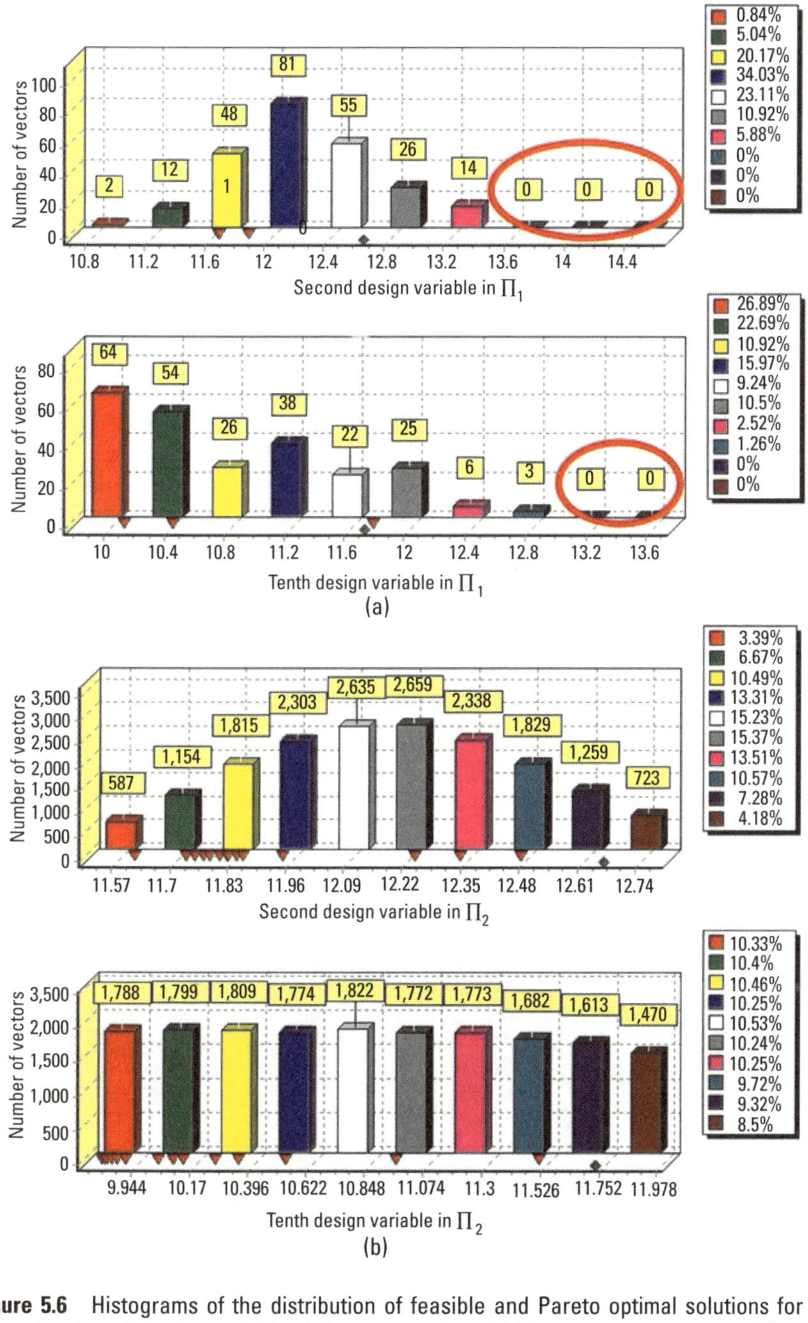

Figure 5.6 Histograms of the distribution of feasible and Pareto optimal solutions for the second and tenth design variables: (a) corresponds to initial parallelepiped Π_1 and (b) corresponds to parallelepiped Π_2. The percentage of designs entering the corresponding interval is indicated on the right of each histogram. The prototype is marked with a green diamond. The Pareto optimal vectors are marked with red triangles. The "gaps" of the initial range for the second and tenth design variables are circled.

- The other boundaries of the parallelepiped Π_2 were determined similarly. In the process of the definition of boundaries of parallelepiped Π_2, the values of the prototype design variables and Pareto optimal solutions #26087, #75527, and #81087 were also considered.

Again, N = 131,072 tests were conducted. N1 = 18,270 vectors entered the test table. The pseudo-criteria constraints were defined based on the analysis of the test tables, while constraints on performance criteria were kept unchanged. This time NF = 17,302 vectors (including the prototype) entered the feasible solution set, and NP = 14 Pareto optimal solutions were identified (#87511, #21855, #63919, #49109, #106467, #90819, #78907, #116871, #31819, #64407, #74223, #80159, #105823, and #22671). It follows that the coefficient of the efficiency of searching the feasible solutions (γ_F) was increased more than 70 times. Furthermore, the histograms in Π_2 have much better distributions of the feasible solutions than in Π_1 [see Figure 5.6(b)]. Table 5.2 shows 3 out of 14 Pareto optimal solutions.

Figure 5.7 shows the dependency between the criterion Φ_1 and the criterion Φ_2 and the new feasible and Pareto optimal solutions. Recall that in the initial statement there were only NF = 240 and NP = 3 feasible and Pareto optimal solutions, respectively. A reader may compare Figure 5.7 with Figure 5.1 for the initial statement of the problem.

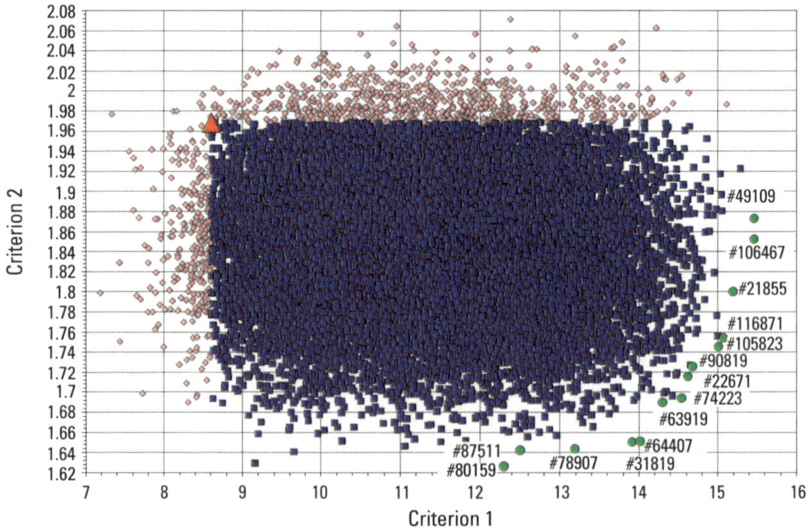

Figure 5.7 The dependency between criteria Φ_1 and Φ_2. The second statement: feasible solutions ND = 17,302; Pareto optimal solutions NP = 14.

After combining the feasible sets in both problem statements, we have constructed a combined Pareto set. No solution belonging to the initial statement has entered into the combined Pareto set. In other words, all the results of optimization in the initial statement have been improved. The preference of the expert was given to the vector #74223. This vector surpassed all solutions in the initial statement by two performance criteria simultaneously (see Table 5.2 and Figure 5.7).

5.1.1.4 Final (Third) Statement and Solution of the Optimization Problem: Parallelepiped Π_3

By analyzing the feasible solutions from Π_2, a new parallelepiped Π_3 was constructed. After N = 131,072 tests, N1 = 34,986 vectors entered the test table. Compared to the initial and second statements of the problem, all criteria constraints now were determined on the basis of analysis of the test tables. In particular, more rigid criteria constraints (Φ_1^{**} = 14.342 and Φ_2^{**} = 1.711) were formulated. The pseudo-criteria constraints were also revised. As a results, we obtained NF = 847 feasible solutions and NP = 7 Pareto optimal solutions.

The smaller number of the feasible solutions compared to the previous problem statement can be explained by much stronger performance criteria constraints. The Pareto optimal solutions (#113487, #4145, #68410, #39801, #53988, #72461, #75110) are shown in Figure 5.8. The values of criteria of these solutions are given in Table 5.2.

The analysis of Pareto optimal solutions revealed that solution #75110 surpassed seven solutions from the second statement (#80159, #87511, #78907, #31819, #64407, #63919, #74223) by two criteria simultaneously. Solution #113487 surpasses five solutions from the second statement (#22671, #90819, #105823, #116871, #21855) by two criteria simultaneously. The expert's preference was given to solution #113487.

Using the feasible solutions from the second and final statement, the combined Pareto set was constructed. The combined Pareto set includes all seven Pareto optimal solutions from the final statement and only two solutions #106467, #49109 belonging to the second statement (see Figure 5.9). Further attempts to improve the obtained solutions have not yielded any new interesting results.

The stability of the most interesting Pareto optimal solutions was investigated with respect to small variations of the parameters in the vicinity of these solutions. To this end, we constructed parallelepipeds centered in the Pareto optimal solutions and performed 1,024 tests in each parallelepiped. The corresponding variations in the criteria were small and insignificant, which indicated the stability of the solutions.

The overall dynamics of improving a prototype on the basis of two corrections of the problem statement is shown in Figure 5.9.

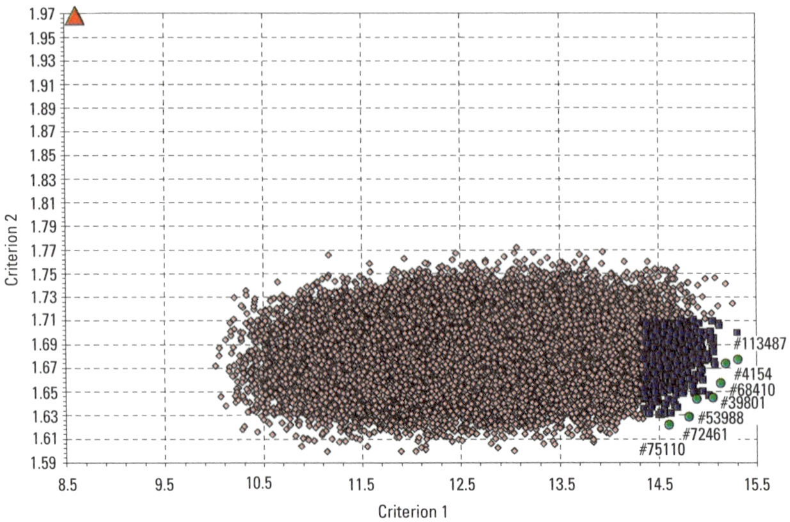

Figure 5.8 Dependency between criteria Φ_1 and Φ_2. Final statement: feasible solutions ND = 847; Pareto optimal solutions NP = 7.

In summary, the problem of improving a preliminary ship design prototype has been solved. To achieve the goal, we corrected the problem statement twice. Each time, analyzing the results allowed us to formulate the improved statement. This process included changing the range of variation of design variables and revising all criteria constraints. As a result of multicriteria analysis, the parallelepiped Π_2 was constructed. The volume of parallelepiped Π_2 was considerably decreased in comparison with the volume of parallelepiped Π_1. Similarly, the parallelepiped Π_3 was constructed with a smaller volume than of parallelepiped Π_2. This allowed a more careful and efficient investigation of the design variable and criteria spaces.

5.2 Problem with the High Dimensionality of the Design Variable Vector

Multicriteria analysis of a ship design model using the PSI method was first performed by O. M. Berezanskii and Y. N. Semenov from the State Sea Technical University, Saint Petersburg [21]. The study was intended to improve the performance criteria of a prototype ship, UT-704, which was built by the ULSTEIN Group in Norway for the oil and gas industry fleet. The performance criteria were tonnage, speed, capital investments, and operational costs. The statistical data on this type of ship was used to choose the design variable constraints. As a result, the prototype has been considerably improved.

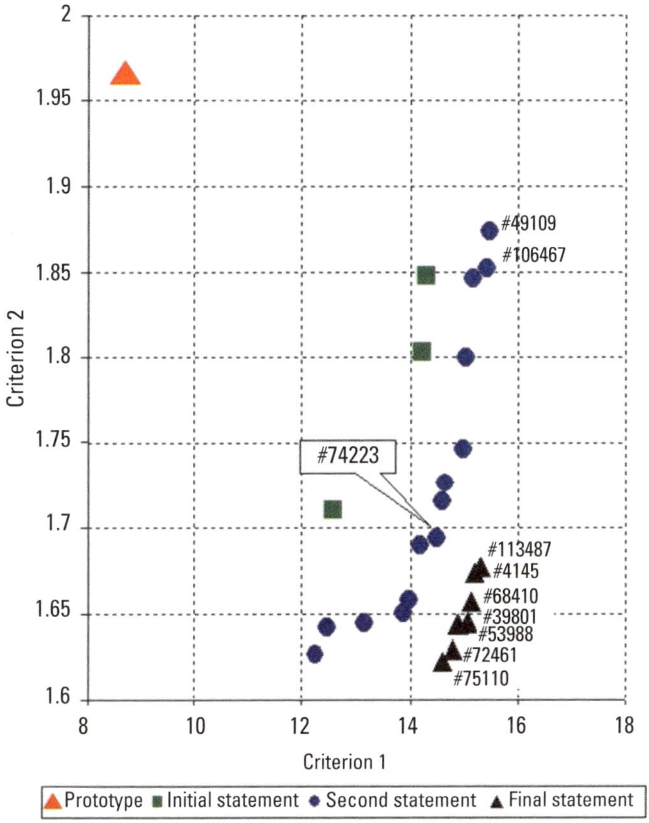

Figure 5.9 Pareto optimal solutions in the three statements and the combined Pareto optimal solutions.

Consider a problem of improving the prototype with high dimensionality of the design variable vector that is described in detail in [2, 14, 15]. Among the particular features of the problem there are the high dimensionality of the design variable vector (45 design variables) and the difficulties of improving a reasonably good prototype under strong constraints (7 functional constraints and 15 performance criteria and pseudo-criteria). Six criteria were optimized: Φ_1 is the propulsion power factor (%) (min); Φ_2 is the electrical power factor (%) (min); Φ_3 is the volume factor (%) (max), Φ_4 is the region factor (%) (max), Φ_5 is the weight factor (%) (max), and Φ_6 is the cost (min). After solving the initial problem, we repeatedly corrected constraints. We refer to these investigations as experiments. Since the purpose of this section is to show the iterative process of the initial data correction, we will omit a description of the mathematical model and briefly illustrate some elements of multicriteria analysis.

In view of the high dimensionality of the design variable vector, 200,000 tests were conducted in each of the first five experiments[3] and 500,000 tests were conducted in the sixth experiment.

Each subsequent experiment was carried out on the basis of the previous one (step by step). In the first two experiments (in parallelepipeds Π_1 and Π_2), no feasible solutions were found; in the third experiment (parallelepiped Π_3), a few feasible solutions were obtained: #17311, #108455, and #71279. These solutions attracted the attention of the expert. For example, design #108455 proved to be better than the prototype in five of the six criteria. Some dependencies between criteria (third experiment) are shown in Figures 5.10 through

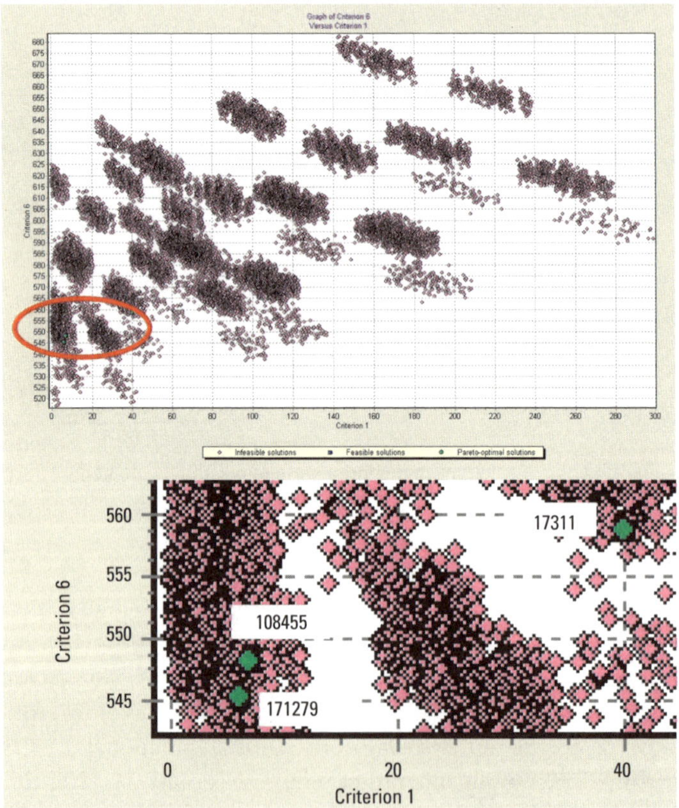

Figure 5.10 Criterion 1 versus criterion 6: third experiment (the scale is increased in the lower sub-figure). Each point has a dimensionality equal to 6 in the criteria space and equal to 45 in the design variable space. Three Pareto optimal solutions (#108455, #171279, and #17311) have been obtained after 200,000 trails. Pareto optimal solutions are green. Unfeasible solutions are magenta.

3. LP_τ sequences were used in experiments 1–4 and 6, and a random number generator in experiment 5. We will restrict ourselves to describing the experiment with the LP_τ sequences.

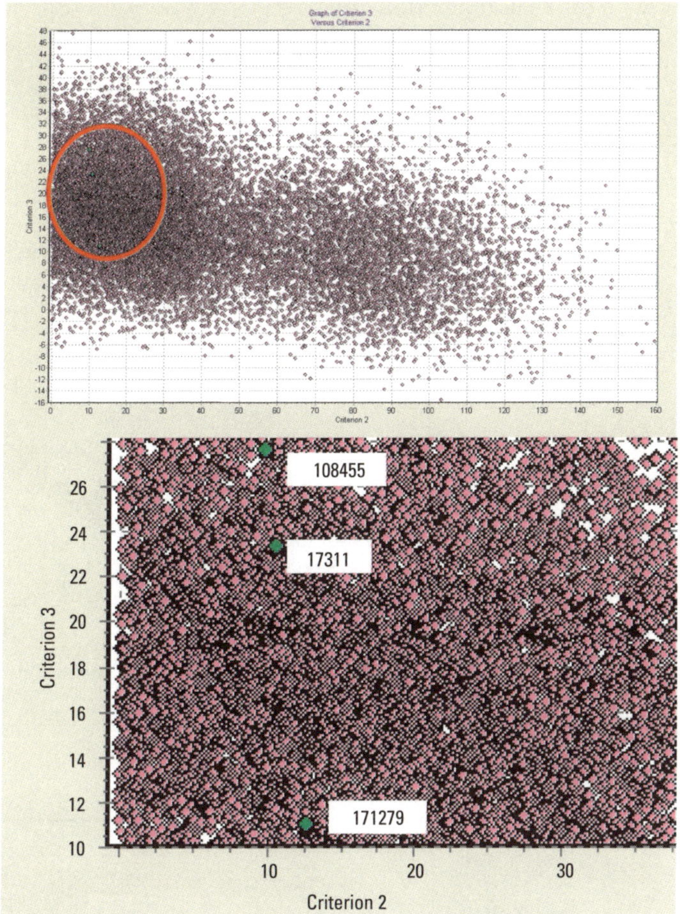

Figure 5.11 Criterion 2 versus criterion 3: third experiment (the scale is increased in the lower sub-figure). Pareto optimal solutions are green. Unfeasible solutions are magenta.

5.13. The regions of the three specified designs are circled. Based on the results of an analysis of the third experiment, the search region in the fourth experiment was limited by the design variable values of the three specified designs. Thus, parallelepiped Π_4 was constructed. In the fourth experiment, the criteria constraints were strengthened in comparison with the third experiment (the first dialogue of the expert with the computer). The number of the feasible and Pareto optimal solutions turned out to be rather high (2,161 and 208, respectively). This is due to the fact that the search region in parallelepiped Π_4 was substantially smaller than in parallelepiped Π_3 for the same number of tests. Many solutions of interest to the expert were found. After analyzing the obtained solutions, an attempt was made to improve the prototype in all criteria

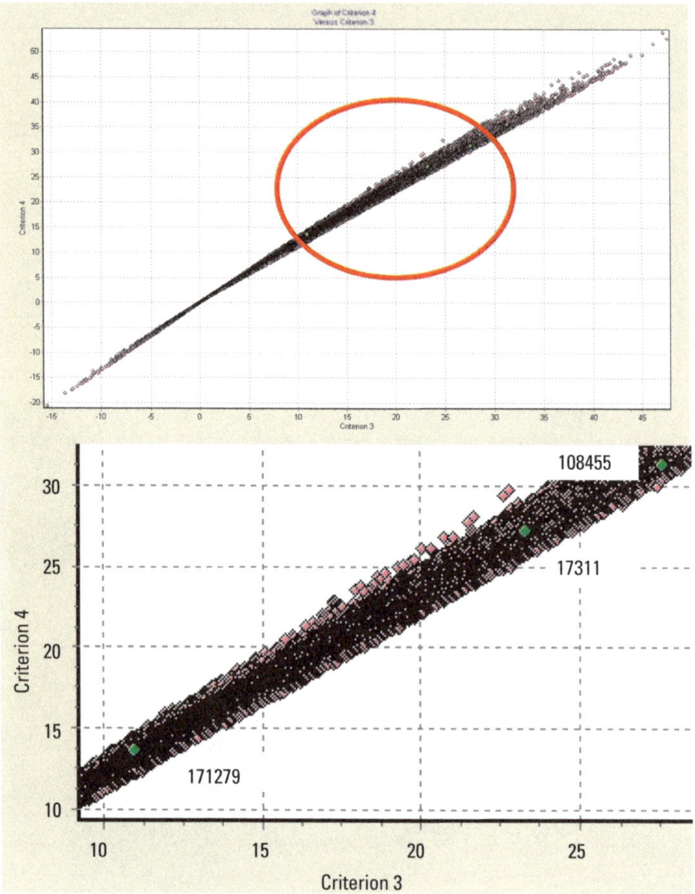

Figure 5.12 Criterion 3 versus criterion 4: third experiment (the scale is increased in the lower sub-figure). Pareto optimal solutions are green. Unfeasible solutions are magenta.

simultaneously. Therefore, in the second dialogue, the criteria constraints corresponded to the values of the prototype criteria. As a result, 20 Pareto optimal solutions surpassing the prototype in all criteria were found. The improvement of the feasible solution set at the expense of correcting constraints is demonstrated in Figures 5.14 through 5.17 (fourth experiment). The analysis of these dependencies and the location of the feasible solution set has been of significant interest for experts.

In the sixth experiment (parallelepiped Π_4) criteria constraints on the second and sixth criteria were strengthened in comparison with the first dialogue in the fourth experiment. A total of 500,000 trials were conducted, and 627 feasible and 138 Pareto optimal solutions, respectively, were found. Six Pareto optimal solutions surpassing the prototype in all criteria were found.

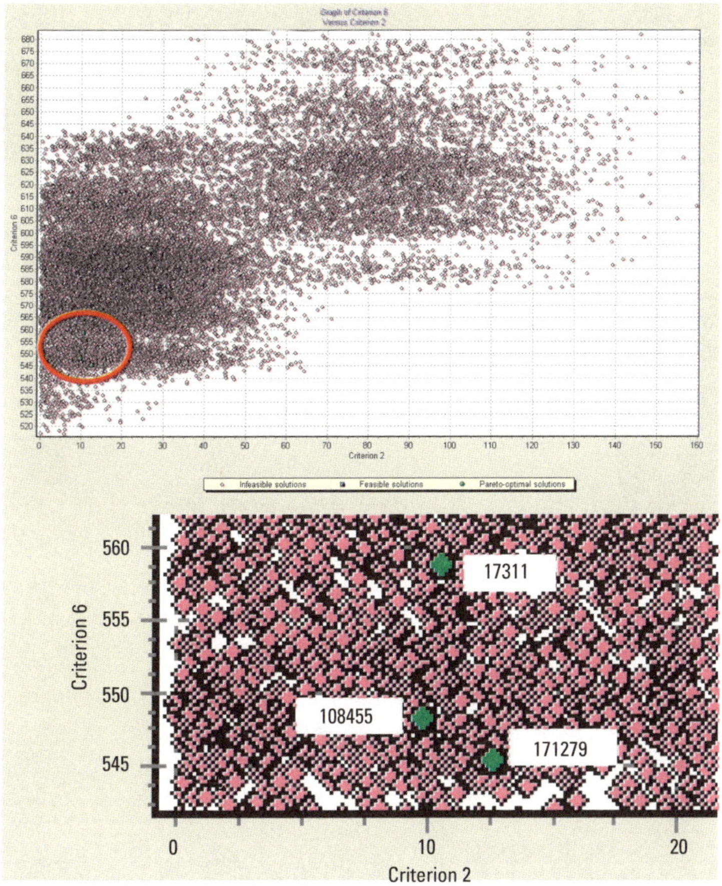

Figure 5.13 Criterion 2 versus criterion 6: third experiment (the scale is increased in the lower sub-figure). Pareto optimal solutions are green. Unfeasible solutions are magenta.

A combined set of Pareto optimal solutions surpassing the prototype in all six criteria contains 26 solutions, five of which are given in Table 5.3.

In summary, we draw attention once again to some features and strategies of problem solving:

- The dimensionality of the design variable vector was high. Therefore, it was necessary to carry out a large number of trials.
- Multicriteria analysis showed the necessity of the repeated correction of the constraints, and, because of this, a series of experiments was performed.
- The prototype was improved by six criteria simultaneously.

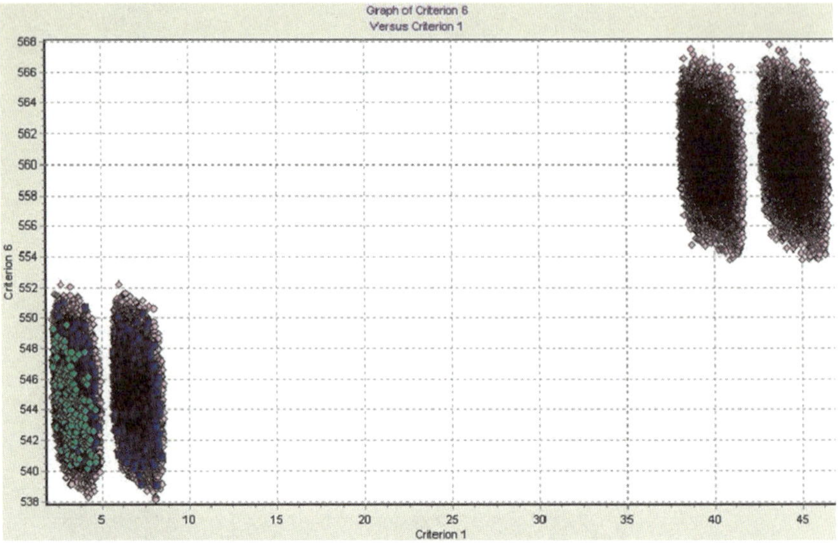

Figure 5.14 Criterion 1 versus criterion 6: fourth experiment. We carried out 200,000 trials. We obtained 2,161 feasible solutions and 208 and Pareto optimal solutions. Feasible solutions are blue and green points; Pareto optimal solutions are green points. Unfeasible solutions are magenta.

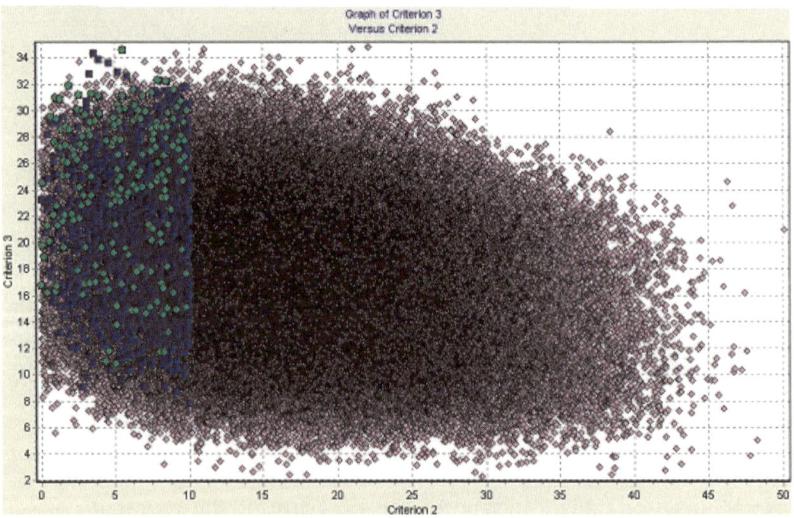

Figure 5.15 Criterion 2 versus criterion 3: fourth experiment. Feasible solutions are blue and green points; Pareto optimal solutions are green points. Unfeasible solutions are magenta.

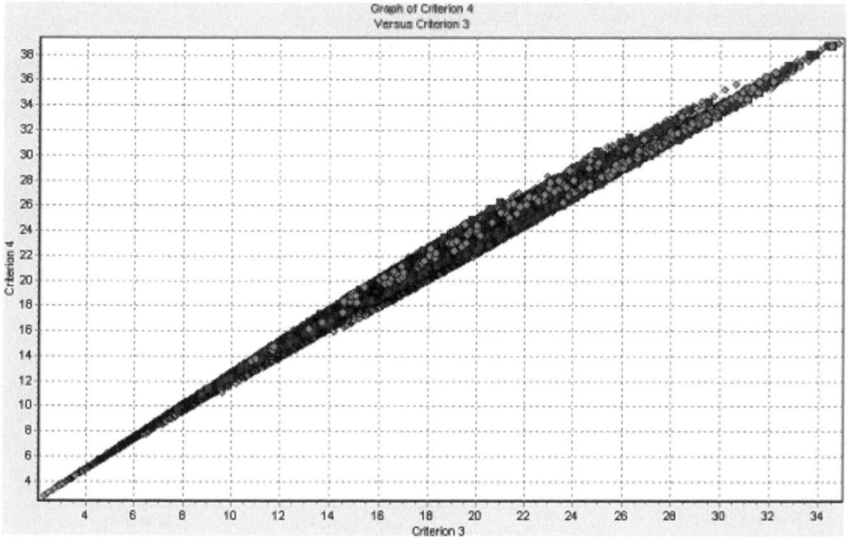

Figure 5.16 Criterion 3 versus criterion 4: fourth experiment. Feasible solutions are blue and green points; Pareto optimal solutions are green points. Unfeasible solutions are magenta.

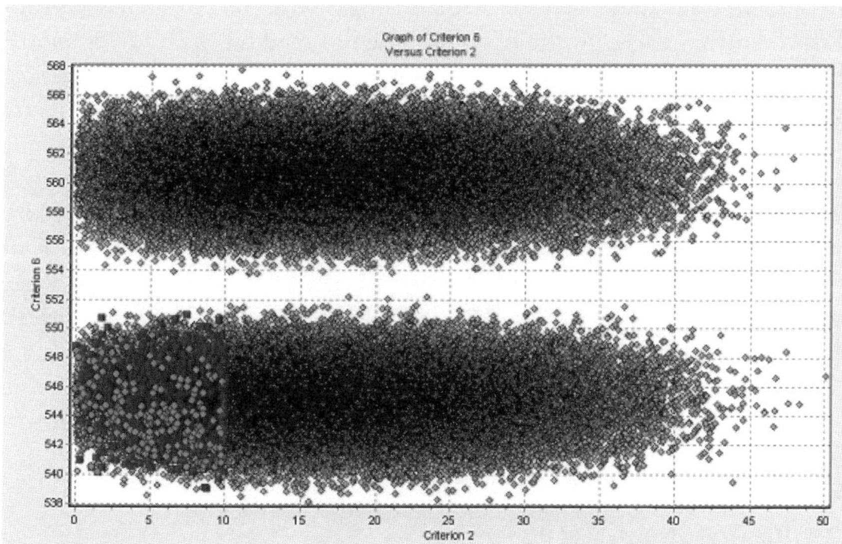

Figure 5.17 Criterion 2 versus criterion 6: fourth experiment. Feasible solutions are blue and green points; Pareto optimal solutions are green points. Unfeasible solutions are magenta.

Table 5.3
Results

Experiments	Φ_1% (min)	Φ_2% (min)	Φ_3% (max)	Φ_4% (max)	Φ_5% (max)	Φ_6 (min)
Prototype	2.48	10.00	11.77	14.33	1.01	555
Fourth experiment, #16907	2.28	8.10	14.6	18.3	5.68	547
Fourth experiment, #164167	2.22	3.03	19.6	23.7	8.46	549
Fourth experiment, #191033	2.42	7.35	15.1	18.7	8.16	544
Sixth experiment, #293036	2.40	1.55	23.4	27.4	2.10	543
Sixth experiment, #364925	2.37	2.56	24.8	28.8	5.14	547

5.3 Rear Axle Housing for a Truck: PSI Method with the Finite Element Method

The application of the PSI method in combination with the finite element method (FEM) is discussed in this section through Section 5.5. The approach of combining the finite element method (FEM) with the PSI method consists of the following stages. At the first stage, the PSI method (with LP_τ or another generator) is used to define N vectors of parameters (N projects). At this stage geometrical parameters and types of materials are most often varied. At the second stage, FEM is used to select N^{FEM} projects ($N^{FEM} \leq N$). N^{FEM} projects should satisfy the standards and other rigid requirements (rigid functional constraints). At the third stage, the PSI method is used to form a test stable for these N^{FEM} projects. The criteria constraints are imposed in the interactive mode on the basis of the analysis of the test table. Then the feasible solutions and the Pareto optimal solutions, NP ($NP \leq N^{FEM}$), are constructed and analyzed, and next the most preferable solution is defined [8].

We briefly consider the rear axle housing for a truck rated at 2.5 tons. The objective of the investigation is to reduce the housing mass and improve its strength and other characteristics.

5.3.1 General Statement of the Problem

We consider static loading of the rear axle housing by the forces transmitted from the rear suspension and drive wheels. By solving the static problem, we determine the equivalent stresses in the structure, the housing displacements, and the angular deviations of the half-axles from the final drive axis [1, 17].

Thc finite element model of the housing is shown in Figure 5.18(a, b). It contains 1,456 elements and 1,238 nodes. In modeling the housing, we used three- and four-node shell elements.

Consider the following problem statement. The *design variables* to be varied are the thicknesses of different parts of the structure:

- α_1 is the thickness of the housing walls;
- α_2 is the thickness of the bearing housings;

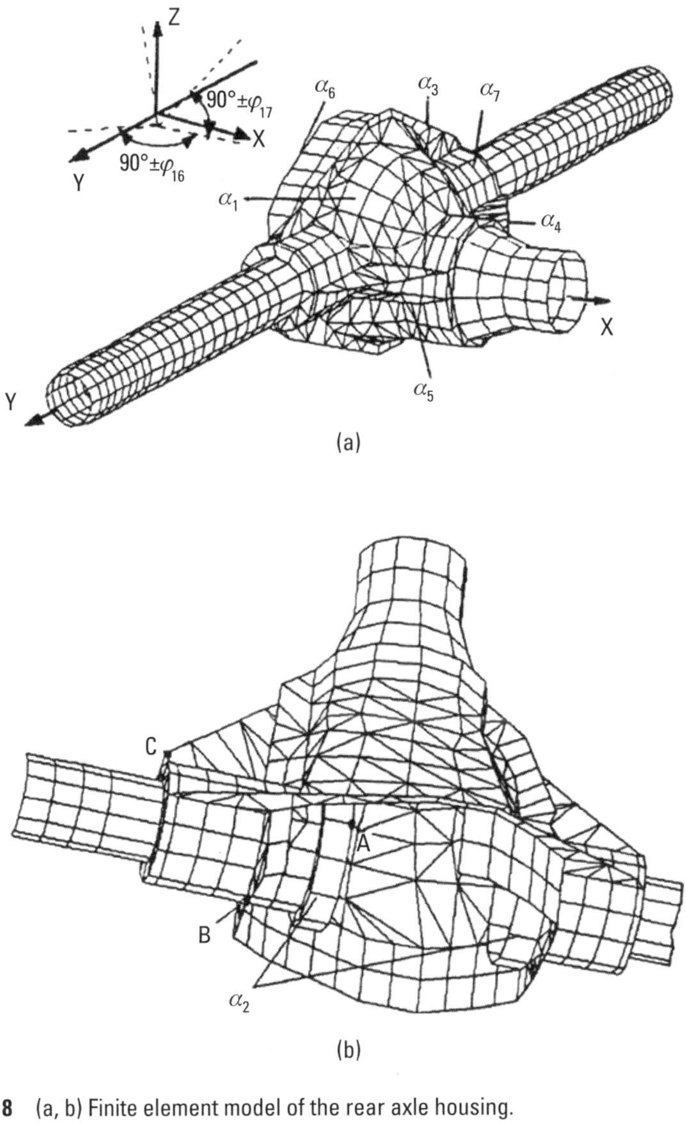

Figure 5.18 (a, b) Finite element model of the rear axle housing.

- α_3 is the thickness of the vertical stiffening rib;
- α_4 and α_5 are the thicknesses of the horizontal stiffening ribs;
- α_6 is the rear cover of the axle housing;
- α_7 is the half-axle housing bush.

The thicknesses $\alpha_1 - \alpha_7$ are shown in Figure 5.18. In the following, these thicknesses are given in millimeters. The ranges of the design variables are as follows:

$$4 \le \alpha_1 \le 7; \quad 4 \le \alpha_2 \le 13; \quad 4 \le \alpha_3 \le 13; \quad 4 \le \alpha_4 \le 11;$$
$$4 \le \alpha_5 \le 11; \quad 4 \le \alpha_6 \le 7; \quad 4 \le \alpha_7 \le 13$$

In our case the *functional dependences* $f_l(\alpha)$ are the equivalent stresses in the structure. These stresses are investigated at the 12 most dangerous places of the structure, and thus $l = \overline{1,12}$. The housing material is high-strength cast iron. The maximum allowable value of the equivalent stress calculated according to the Huber-Mises theory was set at $[\sigma]_{max} = 15$ kgf/mm^2 (147 N/mm^2), and thus the functional constraints are

$$f_l(\alpha) \le 15, \quad l = \overline{1,12}$$

The *performance criteria* to be optimized involve:

- The total mass of the structure Φ_1 (kg);
- The maximum deflection in the structure Φ_2 (mm);
- The maximum equivalent stresses in the structure Φ_3 (kgf/mm^2), which must not exceed the maximum allowable value $[\sigma]_{max}$.

Since we are dealing with a mass production of trucks, it is desirable to minimize the rear axle housing mass. The reduction in the structure deflection increases its durability. Moreover, it is necessary to minimize the peak equivalent stresses.

Two additional performance criteria, Φ_{16} and Φ_{17} (in degrees), characterize the change (as a result of the housing deformation) in the angular positions of the half-axles with respect to the final drive axis. These two criteria should also be minimized. This would improve the working conditions for the differential by reducing tooth wear and increasing the contact and bending endurance of the teeth. All five criteria also characterize the working conditions of the bearings.

In order to solve this problem, two approaches are possible. The first approach is to solve the problem with five performance criteria (Φ_1, Φ_2, Φ_3, Φ_{16}, and Φ_{17}) and 12 rigid functional constraints.

As has already been said, the more criteria are optimized, the more information the expert obtains about the work of the mathematical model. In connection with this, we represented all functional relations as pseudo-criteria. Thus, the second approach is to solve the problem with criteria and pseudo-criteria. According to what has been said earlier, instead of the functional relations $f_l(\alpha)$, $l=\overline{1,12}$, which describe the equivalent stresses in the 12 most dangerous points of the structure, we introduce pseudo-criteria $\Phi_4,\ldots,\Phi_{15}$. The value of Φ_3 is determined as the maximum over the values $\Phi_4, \ldots, \Phi_{15}$ for some fixed design variable vectors α^i, $i=\overline{1,N}$, $\Phi_3(\alpha) = \max\limits_{4\le v\le 15} \Phi_v(\alpha)$. Three of the 12 points, A, B, and C, are shown in Figure 5.18(b). The pseudo-criteria $\Phi_4(\alpha)$, $\Phi_7(\alpha)$, and $\Phi_{13}(\alpha)$ determine the equivalent stress values at the points A, B, and C, respectively. As will be shown next, the stress is a maximum at these points.

Thus, the vector of criteria contains Φ_v, $v=\overline{1,17}$.

We preferred the second approach for following reasons. In process of the analysis of test tables, the expert obtains interesting information about distributions of stresses at the 12 most dangerous places, depending on the housing geometry (wall thickness). It allows us to define how much these points are dangerous in reality.

5.3.2 Solution of the Problem and Analysis of the Results

We carried out 256 trials. Finally, the following criteria constraints were adopted:

$$\Phi_1^{**}=71.5;\Phi_2^{**}=1.33;\Phi_3^{**}=\Phi_4^{**}=\ldots=\Phi_{15}^{**}=15;\Phi_{16}^{**}=0.031;\Phi_{17}^{**}=0.031$$

The feasible set comprised 22 designs, of which 14 were Pareto optimal designs: 21, 43, 62, 67, 79, 110, 113, 158, 174, 181, 229, 233, 246, and 254.

Table 5.4 (the unordered test table) gives the values of the performance criteria. For designs 38, 62, 110, 129, 158, 174, 206, 233, and 242 the stress attains its maximum value at point A (pseudo-criterion Φ_4), where the bearing is located [see Figure 5.18(b)]. The record 13.45 (A) in the column Φ_3 of Table 5.4 means that for design 38 the maximum stress is equal to 13.45 kgf/mm^2 (point A).

For designs 21, 23, 43, 77, 79, 113, 121, 181, 189, 229, 246, and 254, the stresses attain their maximum values at point B (pseudo-criterion Φ_7), at the bush-housing contact [see Figure 5.18(b)]. In the column Φ_3 of Table 5.4, the record 13.11 (B) means that to design 21 there corresponds a maximum stress equal to 13.11 kgf/mm^2 at point B.

Table 5.4
Feasible and Pareto Optimal Solutions

Φ^i	Φ_1	Φ_2	Φ_3	Φ_{16}	Φ_{17}
21	70.31	1.31	13.11(B)	0.022	0.026
23	71.42	1.29	14.57(B)	0.021	0.024
38	70.99	1.26	13.45(A)	0.025	0.029
43	71.41	1.27	14.13(B)	0.018	0.022
62	70.31	1.24	12.28(A)	0.019	0.024
67	70.64	1.26	11.50(C)	0.024	0.027
77	71.21	1.29	14.15(B)	0.022	0.026
79	70.71	1.27	12.68(B)	0.018	0.023
110	70.70	1.23	12.43(A)	0.021	0.026
113	69.38	1.31	14.43(B)	0.023	0.026
121	70.53	1.29	13.28(B)	0.025	0.029
129	70.66	1.30	14.85(A)	0.027	0.029
158	70.37	1.26	13.70(A)	0.019	0.024
174	70.31	1.27	13.41(A)	0.024	0.028
181	70.15	1.27	13.42(B)	0.022	0.025
189	70.78	1.33	14.52(B)	0.026	0.028
206	71.24	1.23	13.68(A)	0.022	0.026
229	70.04	1.27	12.67(B)	0.023	0.027
233	69.85	1.27	11.53(A)	0.023	0.027
242	69.91	1.28	14.42(A)	0.024	0.029
246	68.96	1.32	14.41(B)	0.026	0.028
254	69.63	1.32	13.99(B)	0.029	0.031

It follows from Table 5.4 that, of all the feasible designs, only design 67 has the maximum stress at point C (the horizontal stiffening rib, pseudo-criterion Φ_{13}) [see Figure 5.18(b)]. This stress is equal to 11.50 kgf/mm^2.

Pareto optimal designs 67 and 233, which are characterized by comparatively low maximum stresses, were of particular interest to the experts. Moreover, design 233 is one of the best with respect to mass and deflection. It is not surprising that design 233 was chosen as the most preferable solution. For this solution the design variable vector is α^{233} = (5.77; 12.19; 4.46; 7.85; 4.3; 4.27; 6.07).

The antagonism of criteria Φ_1 and Φ_2 is obvious from Figure 5.19.

In order to visualize the action of stresses (pseudo-criteria) in the most dangerous places of the housing, we present Figure 5.20. This figure shows

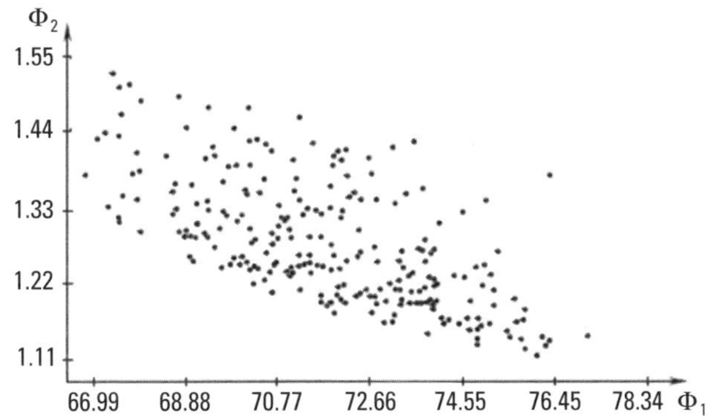

Figure 5.19 Projections of 256 criteria vectors onto plane Φ_1-Φ_2.

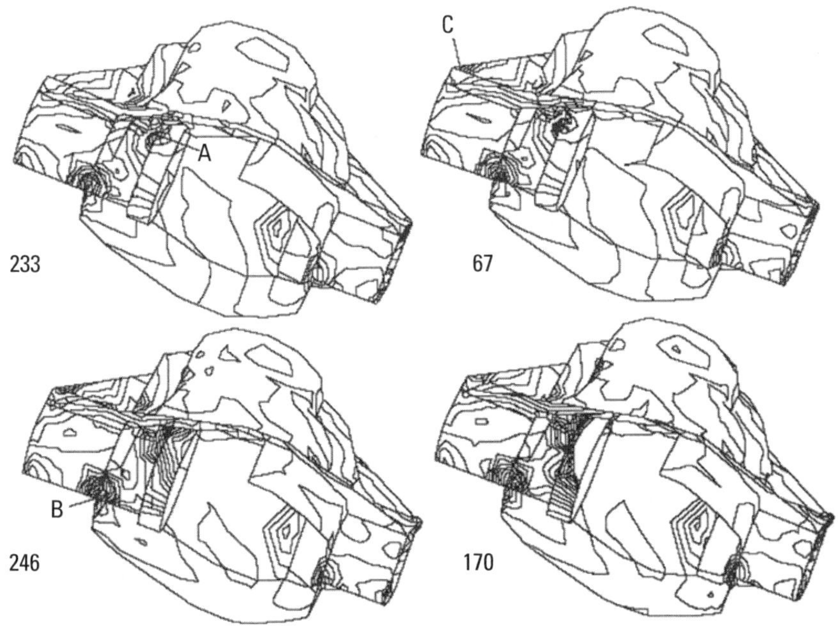

Figure 5.20 Visualization of the action of stresses (taken as pseudo-criteria) at the most dangerous places of the housing.

the stress level lines for the Pareto optimal designs 67, 233, and 246. For comparison with these designs, we indicate the unfeasible design 170, in which the housing mass is comparatively small, Φ_1 = 67.52 kg, whereas the deflection is large, Φ_2 = 1.55 mm. This design does not meet criteria constraints

Φ_2^{**}, Φ_7^{**}, and Φ_{16}^{**}. The stresses at point A are very high and are practically equal to the limiting values, and at point B they are even higher. The stresses at points A, B, and C are equal to 14.98, 16.86, and 9.93 kgf/mm^2, respectively.

The stress level lines for unfeasible design 170 are shown in Figure 5.20. For this solution, the design variable vector is α^{170} = (4.99; 4.10; 6.63; 8.78; 5.33; 5.18; 5.16) and the criterion vector is $\Phi(\alpha^{170})$ = (67.52; 1.55; 16.86; 0.032; 0.027).

5.3.3 Conclusions

After analyzing the results obtained, taking into account the housing's mass, maximum deflections, and maximum stresses at various places in the structure, and the change in the angular position of the half-axles with respect to the final drive axis, the experts preferred design 233, as was mentioned earlier. This design is characterized by practically identical and comparatively low stresses at the most dangerous places of the structure.

Compared with the prototype, the design was improved in all criteria; in particular, the mass was reduced by 5 kg and the maximum deflection was reduced by 15%. Note that the optimal design 233 is the best for neither of the particular criteria. Nevertheless, the experts unanimously adopted design 233 as the optimal solution, and this choice caused no difficulties.

In many engineering structures, including machines, we encounter a typical situation where there are several very highly loaded places in which stresses are high and damage is possible. In the other places, which make up the majority, the stresses are substantially lower. The analysis of the pseudo-criterion values shows that the maximum stresses occur in the housing points that are marked by the letters A, B, and C in Figure 5.18(b). For the feasible designs 23, 113, 129, 189, 242, and 246, the maximum stresses at points A and B approach the maximum allowable value of 15 kgf/mm^2.

5.4 Improving the Truck Frame Prototype

As the major structure of a truck, the frame is subjected to the influences of both the roughness of the road and the units mounted on the truck itself. In turn, the properties of the frame strongly affect many significant characteristics of a truck, such as its controllability, smoothness of motion, vibration loads, stability, and so forth. Furthermore, the mass of the frame makes up a considerable portion of the overall mass of a truck.

A frame is designed subject to conflicting requirements. One has to decrease its mass and at the same time enhance its strength and ensure the specified level of a number of operational characteristics.

Here we consider the problem of designing an optimal truck frame, formulated as follows: It is necessary to design a frame whose mass is smaller than that of the prototype and whose strength properties are improved compared with the latter. In addition, the stiffness characteristics of the optimal frame must be close to those of a prototype whose dynamic properties are sufficiently high.

5.4.1 History of This Project

During World War II the United States granted the Soviet Union invaluable help that included various advanced technical equipment and food (by Lend-Lease). Studebaker trucks were specifically involved and appreciated by many. In Russia in the 1990s the most popular trucks were ZIL. One of the best ZIL modifications used the Studebaker frame. Since the Russian truck was created to carry less cargo, compared with the U.S. truck, it was necessary to reduce the weight of the frame. The PSI method was used in conjunction with the finite element method (FEM). The latter was presented as the TPS–10 software.

5.4.2 Finite Element Model of a Truck Frame

Figure 5.21 shows a model composed of platelike elements possessing both membrane and flexural stiffness. By using these elements, one can take into account the effects of stiffened torsion in the joints of the frame and in the zones

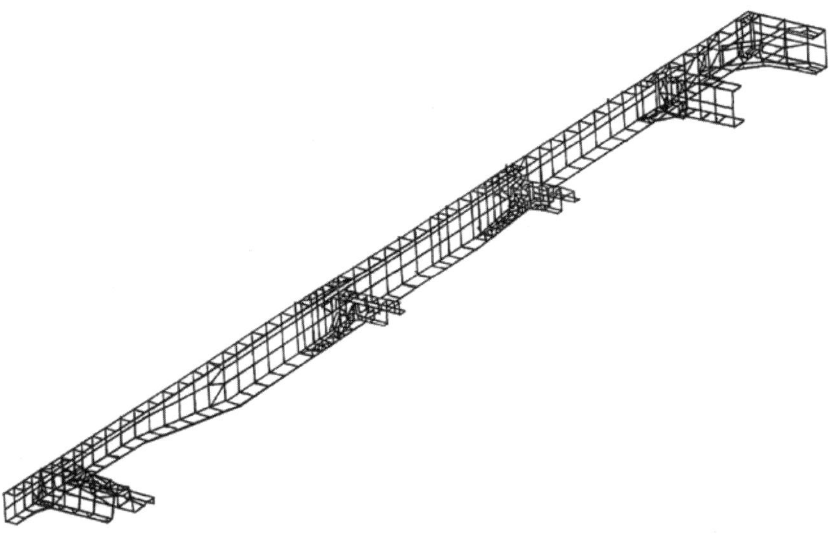

Figure 5.21 Finite element model of the frame.

where the cross beams (cross pieces) are fastened to the longitudinal beam (side rail) in the most natural way and analyze the stressed state of the structure under study in sufficient detail [1, 8].

Numerous bench and road tests confirmed the adequacy of the model. The calculated and experimental results were compared for the major loading modes resulting in torsion and bending in the vertical and horizontal planes. The loading modes were chosen taking into account the statistics of truck frame failures. The model allows estimation of a stress strain state taking into account the specific features of interaction of the frame's elements. It proved to be highly efficient in determining the dynamic characteristics of a truck.

5.4.3 Criteria and Pseudo-Criteria

As previously mentioned, we chose as a prototype the frame that showed good performance in a number of criteria under the long-term operation of the truck. Therefore, the experts wanted to preserve in the new structure the twisting angle of the frame, as well as the bending stiffness in the vertical and horizontal planes, if possible. In other words, we faced the problem of functional relations subjected to "soft" constraints. The experts were ready to revise these constraints, depending on the results achieved. That is precisely why these functional constraints are represented in the form of three pseudo-criteria.

The first pseudo-criterion, $\Phi_1(\alpha)$, is the relative difference between the twisting angles of the frame being designed and those of the prototype.

Let φ_i be the twisting angle of the frame being designed in the ith trial $(i = \overline{1, N})$ and φ_p be the twisting angle of the prototype frame. Then $\Phi_1(\alpha) = (\varphi_i - \varphi_p)/\varphi_p$. Usually, $\Phi_1(\alpha)$ is expressed in percent; in this case, $\Phi_1(\alpha) = [(\varphi_i - \varphi_p)/\varphi_p] \cdot 100\%$ The other pseudo-criteria, $\Phi_2(\alpha)$ and $\Phi_3(\alpha)$, which take into account the difference between the maximum deflections in the horizontal and vertical planes in the frame being designed and in the prototype, are defined in a similar way.

The following four performance criteria were minimized:

- $\Phi_4(\alpha)$ is the longitudinal beam mass (in kilograms).
- $\Phi_5(\alpha)$ is the thickness of the rolled metal sheet from which the longitudinal beam is manufactured (in millimeters). This criterion is simultaneously a design variable.
- $\Phi_6(\alpha)$ and $\Phi_7(\alpha)$ are the maximum stresses under torsion and horizontal bending (in kgf/cm^2).

All criteria are to be minimized. Three hundred trials were carried out.

5.4.4 Design Variables

Twenty-one design variables describing the longitudinal beam geometry and stiffness properties of the cross beam were varied. One of the most important variables is the thickness of the rolled metal sheet from which the longitudinal beam is manufactured. This design variable to a large extent determines the mass of the longitudinal beam, as well as the stiffness characteristics and stresses in the structure.

Earlier, in the example with an oscillator (Chapters 2 through 4), we described in detail the construction and analysis of the histograms, the correction of the initial parallelepiped, and the construction of a new one. In the example of the rear axle housing, we sketched the analysis of the feasible set. We do not consider it worthwhile to describe similar investigations for the example in question, and we confine ourselves to the final results presented in Table 5.5.

In this table, Φ_1–Φ_3 are the pseudo-criteria, where the + and − signs stand, respectively, for increasing and decreasing deviations (in percent) from the corresponding values for the prototype; Φ_4–Φ_7 are the performance criteria; Φ_5 is the calculated thickness of the rolled metal sheet; and D is the corresponding standard thickness. For D = 5.5 mm, the best designs are 244, 168, 176, 20, and 36; for D = 5.6 mm, the best designs are 74, 2, 252, and 266; and so on. The last row in Table 5.5 presents the performance criteria values for the prototype; the thickness of the sheet is $\Phi_5 = 6.35$ mm.

In conclusion, we note the convenience and simplicity of analyzing the results of solving the problem using the table of feasible solutions. An analysis of this table showed that there are a number of designs that exceed the prototype in all performance criteria.

In the optimal design 168, the mass is reduced by 28 kg compared with that of the prototype (the mass of each longitudinal beam is reduced by 14 kg) and the stress level in the most dangerous places is also decreased. The thickness of the rolled metal sheet is reduced from 6.35 to 5.5 mm, while the width of the longitudinal beam flange is increased within some allowable limits.

5.5 Multicriteria Optimization of Orthotropic Bridges

5.5.1 Introduction and Purposes

Issues such as construction of modern cable stayed and suspension bridges with large spans would be unachievable without using orthotropic deck systems. Moreover, the recent standards recommend using lightweight orthotropic deck in such structures. In construction of these bridges, not only are the simply open orthotropic plates used, but also the complex orthotropic closed form. Due to the heavy weight of complex orthotropic closed form comparable to the open one, the optimization of the closed form becomes essential to reduce its

Table 5.5
Table of Feasible Solutions

D (mm)	Designs	Φ_1 (%)	Φ_2 (%)	Φ_3 (%)	Φ_4 (kg)	Φ_5 (mm)	Φ_6 (kgf/cm²)	Φ_7 (kgf/cm²)
5.5	244	−8.9	+5.2	−8.2	95.5	5.53	936	1,814
	168	−3.5	+2.7	+4	90	5.46	934	2,046
	176	−4.4	+9.32	−7.5	94	5.44	897	1,824
	20	−1.8	+4.7	+11	92	5.51	883	2,095
	56	+6.48	+7.16	+3.03	91.2	5.48	925	2,067
5.6	74	−6	+3.64	−4	95.3	5.62	933	1,817
	2	−5.7	+8.4	−6.9	94.7	5.57	903	1,947
	252	−4.6	+7.6	−0.7	92.4	5.57	932	1,955
	266	+3.2	+1.7	+7.5	93.4	5.59	870	2,018
5.7	62	−9.7	+8.2	+2.7	94.6	5.74	950	2,071
	90	−8.4	−0.5	−2.5	94.8	5.65	992	1,912
	142	−5.6	+8.2	−4.6	95.3	5.71	858	1,823
	238	−5.4	+7.7	−1.6	95.1	5.73	927	2,043
	166	−3.7	+5.4	+5.8	93.5	5.68	912	2,080
5.8	89	−9.6	+6.7	−13.4	98.6	5.82	886	1,698
	1	−9	+4.9	−1.3	95.9	5.75	886	2,009
	129	−7.3	+8.7	−14.4	97.7	5.75	892	1,945
	97	+5.03	+2.46	+1.2	97.1	5.77	823	1,969
5.9	13	−8.5	+8.67	−8	99	5.88	880	1,785
	67	−8.8	+6.41	−5.5	98.1	5.93	930	1,920
	37	−4	−1.5	+3.9	97.1	5.85	885	2,050
	53	−4	+7.9	+9.54	95.7	5.87	828	2,070
	45	−3	−0.85	+6.08	97.05	5.89	904	2,086
	195	+0.41	+3.52	+4.58	96.7	5.93	821	2,095
6.0	7	−6.5	+3.84	−3.3	98.9	6.01	872	2,012
Prototype		—	-	-	104	6.35	1,000	2,200

self-weight. This section[4] is concerned with the application of the multicriteria optimization technique for the closed orthotropic deck of Suez Canal Bridge in Egypt.

4. This section was written by Dr. Eng. Mohamed Eltantawy Elmadawy and Dr. Eng. Mohamed Ahmed El Zareef, (Mansoura University, Faculty of Engineering, Structural Engineering Department, Egypt).

The orthotropic closed form was used in the construction of the Suez Canal Bridge in Egypt (Figure 5.22), built in 2001, 45 km south of Port Said; this bridge connects the Sinai Peninsula with Egypt. It was planned that this design would make a significant contribution to the development of the Sinai Peninsula. The bridge belongs to a steel cable-stayed girder bridge with a box section, with a main span length of 404m and a total length of 730m. The navigation clearance under the bridge is 70m.

Due to an increase in the cost of energy all over the world, the steel price goes up from time to time, which significantly raises the construction costs of these bridges. The multicriteria approach [22] considered in this section for the optimization of a complex orthotropic deck of the Suez Canal Bridge takes into account contradictory criteria: the mass and the deflection. The formulation and solution of the multicriteria problem in this study are based on the PSI method.

5.5.2 Mathematical Model and Parameters

The mathematical model and description of the software proposed for optimal structural solutions based on the PSI method were presented in [1, 23]. The program is presented as a book of Microsoft Excel using the Visual Basic for Application (VBA) Language. The Microsoft Excel book consists of large numbers of sheets. Each sheet contains the necessary input information required for the preparation and formation of the model of the closed orthotropic deck analyzed by the software Femap-Nastran. In this study, the application of the developed multicriteria approach is considered for the structural optimization of three-pay deck of the Suez Canal Bridge (Figure 5.23).

Figure 5.22 Suez Canal Bridge in Egypt.

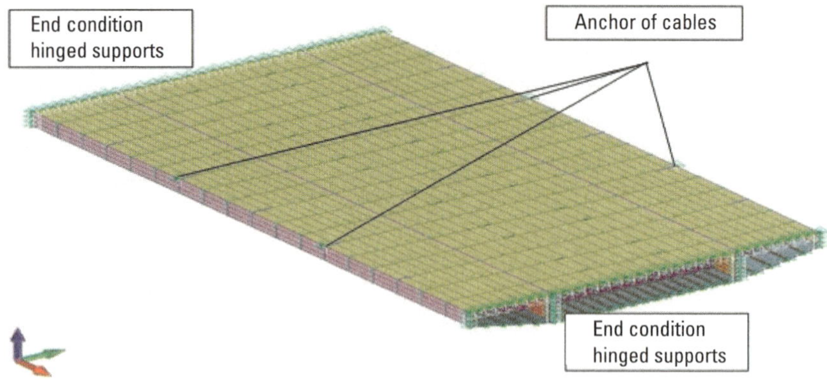

Figure 5.23 Three-pay deck of the Suez Canal Bridge, Egypt.

As shown in Figure 5.23, the physical model consists of a group of vertical and horizontal plates connected to each other and stiffened with longitudinal ribs and transverse beams. The cables are connected to the deck of the bridge at transverse beams. The distance between the anchors of cables in the longitudinal direction of the bridge is 12m, and the distance between the transverse beams is 6m. Two diaphragms are located between the transverse beams and are each 2m.

The model is based on 12 design variables:

- α_1: Thickness of longitudinal ribs, m;
- α_2: Thickness of the transverse beams, m;
- α_3: Thickness of the upper plate, m;
- α_4: Thickness of the lower plate, m;
- α_5: Thickness of the outer vertical plates, m;
- α_6: Thickness of the inner vertical plates, m;
- α_7–α_{12}: The number of longitudinal ribs, which varies in different parts of the structure (1 & 3), (2), (4 & 8), (5 & 7), (6), (9 & 10), as shown in Figure 5.24.

Values of the discrete design variables are shown in Table 5.6.

The previous design variables were used as input parameters for the LP_τ generator that produced 500 possible combinations of these parameters. Some of these obtained combinations are presented in Table 5.7. This process is indicated as stage I-input data as shown in the flowchart in Figure 5.25.

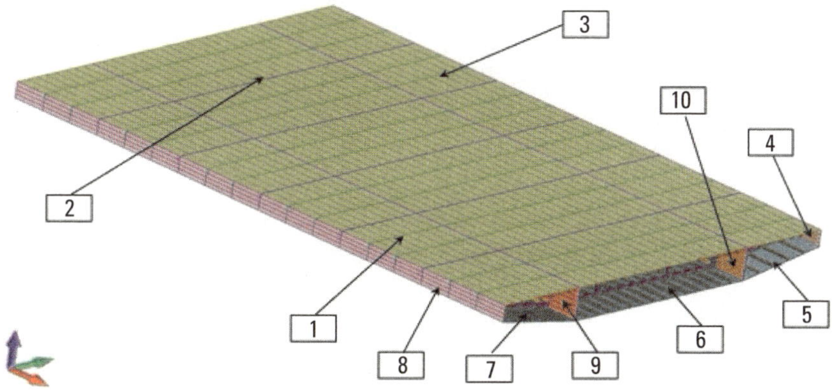

Figure 5.24 Different parts of the Suez Canal Bridge deck.

Table 5.6
Limitations of Design Variables

Thickness of Longitudinal Ribs (m)	Thickness of the Transverse Beams (m)	Thickness of the Upper Plate (m)	Thickness of the Lower Plate (m)	Thickness of the Outer Vertical Plates (m)	Thickness of the Inner Vertical Plates (m)
α_1	α_2	α_3	α_4	α_5	α_6
0.006	0.008	0.012	0.01	0.01	0.01
0.008	0.010	0.014	0.012	0.012	0.012
	0.012	0.016	0.014	0.014	0.014
	0.014		0.016	0.016	0.016
Number of longitudinal ribs (see Figure 5.24)					
Part (1)&(3)	**Part (2)**	**Part (4)&(8)**	**Part (5)&(7)**	**Part (6)**	**Part (9)&(10)**
α_7	α_8	α_9	α_{10}	α_{11}	α_{12}
5	12	2	5	12	4
6	13	3	6	13	5
7	14		7	14	6
	15			15	
	16			16	
	17			17	

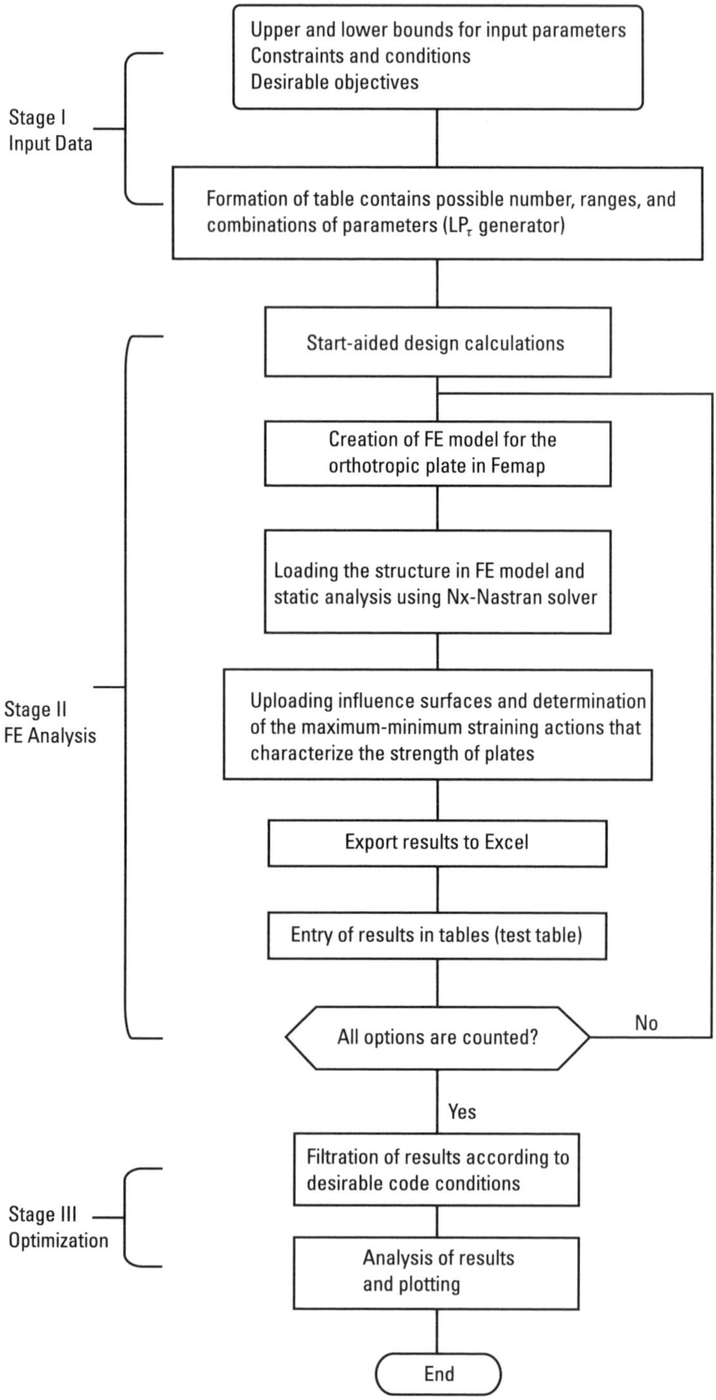

Figure 5.25 Flowchart of the multicriteria optimization technique.

The previous generated combinations were stored in the Microsoft Excel sheet and were sent to the finite element program (FEP) Femap-Nastran using the Visual Basic for Application (VBA) language. The loads considered in the study were dead and live loads. The dead load represents the self-weight of plates, longitudinal ribs, and transverse beams. The vehicles A14 and NK80 were considered according to SNiP 2.05.03-84* [24] as moving live loads.

Through VBA, the main sheet of Microsoft Excel automatically controls the input and output to and from the FEP Femap-Nastran. The obtained results display the values of internal forces and stresses for the longitudinal ribs, the transverse beams, and the plates. The most important output criteria in our case study are:

- Φ_1: The deck self weight of the bridge, kg/m^2;
- Φ_2: The deck deflection, m.

The optimization process aims to minimize both of the deck self-weight (Φ_1) and the deck deflection (Φ_2). Table 5.8 shows all combination parameters and their corresponding output criteria. This stage of solution is indicated as stage II–FE analysis as shown in Figure 5.25.

5.5.3 Results of Optimization

The last stage of the flowchart in Figure 5.25 shows the optimization stage for the results, in which the results are filtered in a separate Microsoft Excel sheet and as a last step in this stage the results will be optimized to get the optimal desirable solution.

In the filtration process, the results will be automatically tested according to SNiP 2.05.03-84* limitations. The 500 combinations and their output criteria are filtered to 161 combinations that achieved the SNiP 2.05.03-84* demands. Some of these filtered results are shown in Table 5.9. In other words, 161 solutions satisfied all functional constraints. One can see that solution #196 has the minimum of self-weight and the maximum of deflection, while solution #445 has the maximum of self-weight and the minimum of deflection. At the end of this stage, the designer should take the decision according to the desirable objectives and field requirements. In the interactive mode the designer imposed the strong criteria constraints $\Phi_1^{**} = 350$ kg/m^2 and $\Phi_2^{**} = 0.0104$ m^2. In consequence of these constraints, four Pareto optimal solutions, #196, #302, #352, and #448, are determined (see Table 5.10). The preference is given to the project #196 that has the minimum of self-weight and the acceptable value of deflection.

Table 5.7
Generated Combinations of Design Variables

Comb. No.	Thickness of Longitudinal Ribs (m)	Thickness of the Transverse Beams (m)	Thickness of the Upper Plate (m)	Thickness of the Lower Plate (m)	Thickness of the Outer Vertical Plates (m)	Thickness of the Inner Vertical Plates (m)
	α_1	α_2	α_3	α_4	α_5	α_6
1	0.008	0.012	0.014	0.014	0.014	0.014
2	0.006	0.012	0.014	0.014	0.012	0.014
3	0.008	0.010	0.016	0.012	0.014	0.012
4	0.006	0.012	0.016	0.016	0.014	0.010
--	--	--	--	--	--	--
498	0.006	0.010	0.016	0.012	0.016	0.014
499	0.008	0.012	0.014	0.016	0.012	0.012
500	0.006	0.010	0.014	0.012	0.012	0.012
	Number of Longitudinal Ribs (see Figure 5.24)					
Comb. No.	**Part (1)&(3)**	**Part (2)**	**Part (4)&(8)**	**Part (5)&(7)**	**Part (6)**	**Part (9)&(10)**
	α_7	α_8	α_9	α_{10}	α_{11}	α_{12}
1	6	15	3	6	15	5
2	6	16	3	6	16	5
3	7	13	2	7	13	6
4	6	14	3	6	15	6
--	--	--	--	--	--	--
498	5	14	3	7	15	4
499	6	16	2	6	13	5
500	5	16	3	5	16	5

5.5.4 Conclusion

The previous study concludes that there is a possibility to make an efficient combination between the PSI method and the finite element program. The efficiency of the new technique is proved through its application to the orthotropic closed deck of the Suez Canal Bridge in Egypt. The optimal solution #196 shows a reduction of 7.2% in weight in comparison with the existing deck, while the deflection is reduced by 1.87%.

Table 5.8
Combination Parameters and Their Corresponding Output Criteria

Comb. No.	Output Criteria: Self weight, (kg/m²) Φ_1	Deflection (m) Φ_2	Thickness (m): Long. ribs α_1	Trans. beams α_2	Upper plate α_3	Lower plate α_4	Outer plates α_5	Inner plates α_6	Number of Longitudinal Ribs: 1, 3 α_7	2 α_8	4, 8 α_9	5, 7 α_{10}	6 α_{11}	9, 10 α_{12}
1	394.682	0.0094135	0.008	0.012	0.014	0.014	0.014	0.014	6	15	3	6	15	5
2	370.695	0.0097712	0.006	0.012	0.014	0.014	0.012	0.014	6	16	3	6	16	5
3	393.497	0.0104051	0.008	0.01	0.016	0.012	0.014	0.012	7	13	2	7	13	6
4	388.835	0.0101709	0.006	0.012	0.016	0.016	0.014	0.01	6	14	3	6	15	6
5	369.466	0.0108866	0.008	0.008	0.014	0.012	0.01	0.014	7	16	2	5	13	5
--	--	--	--	--	--	--	--	--	--	--	--	--	--	--
--	--	--	--	--	--	--	--	--	--	--	--	--	--	--
496	361.329	0.0098724	0.006	0.012	0.014	0.014	0.014	0.012	6	15	2	6	14	5
497	363.051	0.0112991	0.008	0.008	0.012	0.012	0.01	0.016	7	13	3	5	17	6
498	368.979	0.0108648	0.006	0.01	0.016	0.012	0.016	0.014	5	14	3	7	15	4
499	394.402	0.0093522	0.008	0.012	0.014	0.016	0.012	0.012	6	16	2	6	13	5
500	349.629	0.0107724	0.006	0.01	0.014	0.012	0.012	0.012	5	16	3	5	16	5

Table 5.9
Filtered Output Criteria

Comb. No.	Output Criteria: Self weight (kg/m²)	Output Criteria: Deflection (m)	Thickness (m): Long. ribs	Trans. beams	Upper plate	Lower plate	Outer plates	Inner plates	Number of Longitudinal Ribs: 1, 3	2	4, 8	5, 7	6	9, 10
	Φ_1	Φ_2	α_1	α_2	α_3	α_4	α_5	α_6	α_7	α_8	α_9	α_{10}	α_{11}	α_{12}
1	394.682	0.0094135	0.008	0.012	0.014	0.014	0.014	0.014	6	15	3	6	15	5
2	370.695	0.0097712	0.006	0.012	0.014	0.014	0.012	0.014	6	16	3	6	16	5
7	377.116	0.0095092	0.008	0.014	0.012	0.014	0.012	0.012	6	13	3	6	16	4
8	353.829	0.0098537	0.006	0.014	0.014	0.012	0.012	0.01	6	15	3	6	12	4
11	378.751	0.0096558	0.008	0.012	0.014	0.01	0.016	0.012	6	16	3	6	14	6
--	--	--	--	--	--	--	--	--	--	--	--	--	--	--
196	325.156	0.0103033	0.006	0.014	0.012	0.01	0.016	0.012	5	13	2	5	13	6
--	--	--	--	--	--	--	--	--	--	--	--	--	--	--
445	418.413	0.008459	0.008	0.014	0.016	0.016	0.012	0.016	5	15	2	6	16	5
--	--	--	--	--	--	--	--	--	--	--	--	--	--	--
491	374.215	0.0098694	0.008	0.012	0.014	0.014	0.01	0.012	5	13	3	5	13	5
493	377.328	0.0095382	0.008	0.012	0.012	0.014	0.016	0.014	5	15	3	7	13	6
494	391.793	0.0092727	0.006	0.012	0.016	0.016	0.012	0.016	7	16	3	5	14	4
496	361.329	0.0098724	0.006	0.012	0.014	0.014	0.014	0.012	6	15	2	6	14	5
499	394.402	0.0093522	0.008	0.012	0.014	0.016	0.012	0.012	6	16	2	6	13	5

Table 5.10
Pareto Optimal Solutions

Comb. No.	Output Criteria: Self weight, (kg/m^2)	Output Criteria: Deflection (m)	Thickness (m): Long. ribs	Trans. beams	Upper plate	Lower plate	Outer plates	Inner plates	Number of Longitudinal Ribs: 1, 3	2	4, 8	5, 7	6	9, 10
	Φ_1	Φ_2	α_1	α_2	α_3	α_4	α_5	α_6	α_7	α_8	α_9	α_{10}	α_{11}	α_{12}
#196	325.156	0.0103033	0.006	0.014	0.012	0.01	0.016	0.012	5	13	2	5	13	6
#448	330.410	0.0100986	0.006	0.014	0.012	0.01	0.012	0.014	5	12	3	6	14	5
#302	337.648	0.0098049	0.006	0.014	0.012	0.01	0.012	0.014	6	15	2	6	17	4
#352	348.444	0.0096521	0.006	0.014	0.014	0.01	0.012	0.012	5	16	2	6	14	6

References

[1] Statnikov, R. B., and J. B. Matusov, *Multicriteria Analysis in Engineering,* Dordrecht/Boston/London: Kluwer Academic Publishers, 2002.

[2] Statnikov, R. B., et al., *MOVI 1.4 (Multicriteria Optimization and Vector Identification) Software Package,* Certificate of Registration, Registere of Copyright, U.S.A., Registration Number TXU 1-698-418, Date of Registration: May 22, 2010.

[3] Statnikov, R. B., et al., "Multicriteria Analysis Tools in Real-Life Problems," *Journal of Computers & Mathematics with Applications,* Vol. 52, 2006, pp. 1–32.

[4] Statnikov, R. B., A. Bordetsky, and A. Statnikov, "Multicriteria Analysis of Real-Life Engineering Optimization Problems: Statement and Solution," *Nonlinear Analysis,* Vol. 63, 2005, pp. e685–e696.

[5] Statnikov, R., A. Bordetsky, and A. Statnikov, "Management of Constraints in Optimization Problems," *Nonlinear Analysis,* Vol. 71, No. 12, 2009, pp. e967–e971.

[6] Statnikov, R., et al., "Definition of the Feasible Solution Set in Multicriteria Optimization Problems with Continuous, Discrete, and Mixed Design Variables," *Nonlinear Analysis,* Vol. 71, No. 12, 2009, pp. e109–e117.

[7] Xargay, E., et al., "L_1 Adaptive Flight Control System: Systematic Design and V&V of Control Metrics," *AIAA Guidance, Navigation, and Control Conference,* Toronto, Ontario, Canada, August 2–5, 2010.

[8] Statnikov, R. B., and J. B. Matusov, "General-Purpose Finite-Element Programs in Search for Optimal Solutions," *Physics-Doklady (Russian Academy of Sciences),* Vol. 39, No. 6, 1994, pp. 441–443, translated from Doklady Akademii Nauk, Vol. 336, No. 4, 1994, pp. 481–484.

[9] Statnikov, R. B., A. R. Statnikov, and I. V. Yanushkevich, "Permissible Solutions in Engineering Optimization," *Journal of Machinery Manufacture and Reliability,* Russian Academy of Sciences, No. 4, 2005, pp. 1–9.

[10] Bondarenko, M. I., et al., "Construction of Consistent Solutions in Multicriteria Problems of Optimization of Large Systems," *Physics-Doklady,* Vol. 39, No. 4, 1994, pp. 274–279, translated from *Doklady Rossiiskoi Akademii Nauk,* Vol. 335, No. 6, 1994, pp. 719–724.

[11] Lur'e, Z. Y., A. I. Zhernyak, and V. P. Saenko, "Optimization of Pumping Units of Internal Involute Gear Pumps," *Journal of Machinery Manufacture and Reliability,* Russian Academy of Sciences, No. 3, 1996, pp. 29–34.

[12] Podgaets A. R., and W. J. Ockels, "Problem of Pareto-Optimal Control for a High Altitude Energy System," *9th World Renewable Energy Congress WREC IX,* Florence, Italy, August 19–25, 2006.

[13] Podgaets A. R., and W. J. Ockels, "Flight Control of the High Altitude Wind Power System," *Proceedings of the 7th Conference on Sustainable Applications for Tropical Island States,* Cape Canaveral, FL, June 3–6, 2007.

[14] Statnikov, R. B., et al., "Visualization Approaches for the Prototype Improvement Problem," *Journal of Multi-Criteria Decision Analysis,* No. 15, 2008, pp. 45–61.

[15] Anil, K. A., "Multi-Criteria Analysis in Naval Ship Design," Master's Thesis, Naval Postgraduate School, Monterey, CA, 2005.

[16] Statnikov, R., et al., "Visualization Tools for Multicriteria Analysis of the Prototype Improvement Problem," *Proceedings of the First IEEE Symposium on Computational Intelligence in Multi-Criteria Decision-Making (MCDM 2007),* Honolulu, HI, 2007.

[17] Pavlov, Y. S., et al., "Multicriteria Simulation and Analysis," *Journal of Machinery Manufacture and Reliability, Russian Academy of Sciences,* No. 1, 1996, pp. 88–94.

[18] Chernykh, V. V., et al., "Parameter Space Investigation Method in Problems of Passenger Car Design," *Journal of Machinery Manufacture and Reliability, Russian Academy of Sciences,* Vol. 38, No. 4, 2009, pp. 329–334.

[19] Bales N. K., "Optimizing the Seakeeping Performance of Destroyer-Type Hulls," *Proceedings of the 13th ONR Symposium,* Tokyo, Japan, 1980.

[20] Fung, S. C., "Resistance and Powering Prediction for Transom Stern Hull Forms During Early Stage Ship Design," *Department of the Navy, NAVSEA,* Washington, D.C., SNAME Transactions, Vol. 99, 1991, pp. 29–84.

[21] Statnikov, R. B., and J. B. Matusov, *Multicriteria Optimization and Engineering,* New York: Chapman & Hall, 1995.

[22] Demyanushko, I. V., and M. E. Elmadawy, "Application of the Parameter Space Investigation Method for Optimization of Structures," *Transport Construction,* No. 7, 2009, pp. 26–28 (in Russian).

[23] Sobol', I. M., and R. B. Statnikov, *Selecting Optimal Parameters in Multicriteria Problems,* 2nd ed., (in Russian), Moscow: Drofa, 2006.

[24] SNiP 2.05.03-84, "Bridges and Culverts," Moscow, 1988.

6

Multicriteria Identification

One of the major aspects in engineering optimization is the adequacy of the mathematical model to the actual object [1–6]. Without estimating the model's adequacy, the search for optimal design variables has no applied sense, but what is the measure of adequacy? To what extent can we trust one model or another? The central point is the construction of the feasible solution set in multicriteria identification problems. These problems are of great importance for manufacturing motor vehicles, machine tools, ships, aircraft, and other mass-produced objects. In this connection we will also consider problems of the operational development of prototypes. This chapter presents approaches for improving a prototype by the construction of the feasible and Pareto optimal sets while performing multicriteria analysis. Applying the PSI method in multicriteria identification was described in works devoted to an operation development of a vehicle [1], a parafoil-load delivery system [5], a controllable descending system [4], and so on.

6.1 Adequacy of Mathematical Models

Multicriteria identification is a new direction that is of great value in applications. Earlier we solved a number of optimization problems (see Chapter 5). To a considerable extent, our success in this endeavor was owed to the adequacy of the corresponding mathematical models. To have constructed a model of a complex system such that all performance criteria (there may be many dozens of them) are determined with acceptable accuracy is a great success. As a rule, some of the criteria are calculated with comparatively high accuracy, while others are determined with considerable errors. This is the typical situation when investigating complex mathematical models. Therefore, it is very important to have complete information about the mathematical model.

In the most common usage, the term "identification" means the construction of the mathematical model of a system and determination of the parameters (design variables) of the model by using the information about the system response to known external disturbances. In a sense, identification problems are inverse with respect to optimization problems.

When constructing a mathematical model, one first defines the class and the structure of the model operator, that is, the law according to which disturbances (input variables) are transformed into the system response (output processes or the so-called adequacy criteria). This problem is called *structural identification.* For mechanical systems, structural identification means determining the type and number of equations constituting the mathematical model. Structural identification is necessary if there is no preliminary information about the structure of the system or this information is not sufficient for compiling equations. In general, the structural identification problem is very difficult to solve. This apparently accounts for the absence of general methods for solving this problem.

Parametric identification is reduced to finding numerical values of the equation coefficients (parameters of a model), based on the realization of the input and output processes.

The construction of a good (adequate) mathematical model of a complex system is a rare and great success for a researcher. Most often one has to represent the examined system by a set of mathematical models. For example, the planar rigid body model of a truck describes the behavior of the truck at low frequencies fairly well, whereas at high frequencies we have to use a three-dimensional nonlinear model.

When solving optimization problems, we have used the concept of performance criteria. In identification problems we will deal with adequacy (proximity) criteria. By adequacy (proximity) criteria we mean the discrepancies between the experimental and computed data, the latter being determined on the basis of the mathematical model. For example, when identifying the parameters of the dynamical model of a truck, it is necessary to take into account such important particular criteria as vibration accelerations at all characteristic points of the driver's seat, driver's cab, frame, and engine; vertical dynamical reactions at contact areas between the wheels and the road; relative (with respect to the frame) displacements of the cab, wheels, engine, and so forth.

Figure 6.1 shows the location of the characteristic points on a truck at which the experimental values $\mathbf{\Phi}_v^{\text{exp}}$ of the cited criteria should be measured, as well as the arrangement of the vibration exciters. Modern test benches allow real-time simulation of the motion of a truck along roads with various micro-

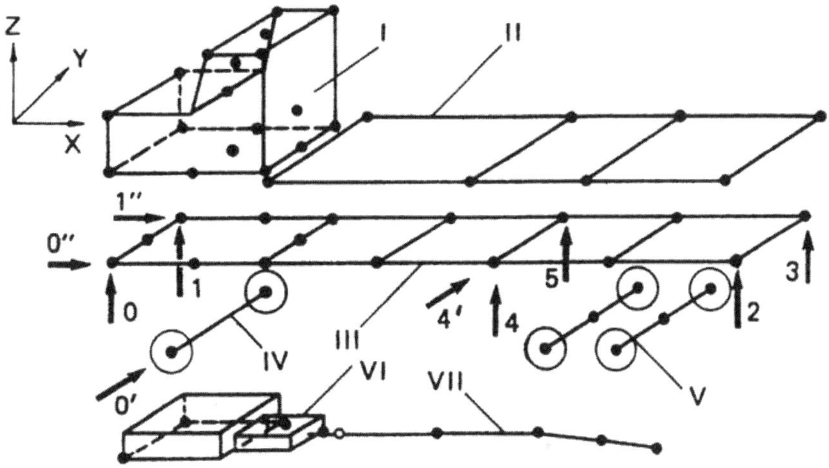

Figure 6.1 Location of the characteristic points on a truck at which the experimental values Φ_v^{exp} are measured. Arrangement of the vibration exciters. I—cab, II—body, III—frame, IV—front axle, V—rear axle, VI—engine, and VII—transmission.

profiles. Points 0–5 correspond to vertical vibrations; 0' and 4' correspond to lateral vibrations; and 0" and 1" correspond to longitudinal vibrations.

The acceleration at the ith point was recorded in the *x*-, *y*-, and *z*-directions. Since the truck design is symmetric, the symmetric, skew-symmetric, torsional, lateral, and longitudinal vibrations are analyzed independently.

In all basic units of the structure under study, we experimentally measure the values of the characteristic quantities of interest (e.g., displacements, velocities, accelerations). At the same time, we calculate the corresponding quantities by using the mathematical model. As a result, particular adequacy (proximity) criteria are formed as functions of the difference between the experimental and computed data. Thus, we arrive at a multicriteria problem. By their nature, identification problems are multicriteria problems. However, as a rule, these problems have been treated as single-criterion problems [7–11]. The multicriteria consideration makes it possible to extend the application area of identification theory substantially. In Section 6.3 the problem of the multicriteria identification of a spindle unit for metal-cutting machines will be considered.

6.2 Multicriteria Identification and Operational Development

Let us discuss some basic features of multicriteria (or vector) identification problems.

1. In the majority of conventional problems, the system is tacitly assumed to be in full agreement with its mathematical model. However, for complex engineering systems (e.g., machines), we generally cannot assert a sufficient correspondence between the model and the object. This does not permit us to use a single criterion to evaluate the adequacy, because it is not corresponding to the physical essence of the problem.
2. Unlike conventional identification approaches, the adequacy of the mathematical model must be evaluated by using a number of particular proximity criteria, as was already mentioned. This multicriteria approach is very important for determining to what extent the mathematical model corresponds to the physical system. For complex systems the number of particular proximity criteria used to evaluate the adequacy of the mathematical model can reach many dozens.
3. Very often, when solving identification problems, the researcher has no information about the limits of many design variables.
4. In structural identification, when we are investigating different mathematical models of the system, the number and limits of the variables to be identified, as well as the number of proximity criteria, can change sign ificantly. Therefore, the problem arises of how to make the identification results for different structures agree.

6.2.1 The PSI Method in Multicriteria Identification Problems

We denote by $\Phi_v^c(\alpha),\ v=\overline{1,k}$ the criteria resulting from the analysis of the mathematical model that describes a physical system, where $\alpha = (\alpha_1,\ldots, \alpha_r)$ is the vector of the model parameters. In problems of mechanics it is often necessary to solve systems of differential or algebraic equations for determining values[1] of criteria $\Phi_v^c(\alpha)$.

Let $\Phi_v^{\exp}$ be the experimental value of the vth criterion measured directly on the prototype. Suppose there exists a mathematical model or a hierarchical set of models describing the system behavior. Let $\Phi = (\|\Phi_1^c - \Phi_1^{\exp}\|,\ldots,\|\Phi_k^c - \Phi_k^{\exp}\|)$, where $\|\cdot\|$ is a particular adequacy (closeness, proximity) criterion. As previously mentioned, this criterion is a function of the difference (error) $\Phi_v^c - \Phi_v^{\exp}$. It is very often given by $(\Phi_v^c - \Phi_v^{\exp})^2$ or $\|\Phi_v^c - \Phi_v^{\exp}\|$.

We formulate the following problem by comparing the experimental and calculation data, determining to what extent the model corresponds to the physical system, and finding the model variables. In other words, it is necessary to find the vectors α^i satisfying the conditions (1.1), (1.2), and the inequalities

1. In problems of medicine and biology we construct $\Phi_v^c(\alpha)$ on the basis of the observations with the use of classification and regression algorithms. See also Section 7.3.

$$\left\|\Phi_v^c\left(\alpha^i\right)-\Phi_v^{\exp}\right\|\le\Phi_v^{**} \tag{6.1}$$

Conditions (1.1), (1.2), and (6.1) define the *feasible solution set* D_α. Here, Φ_v^{**} are criteria constraints that are determined in the dialogue between the researcher and a computer. To a considerable extent, these constraints depend on the accuracy of the experiment and the physical sense of the criteria Φ_v.

Notice that if the experimental values $\Phi_v^{\exp}, v=\overline{1,k}$ are measured with considerable error, then the quantity $\Phi_v^{\exp}$ can be treated as a random variable. If this random variable is normally distributed, the corresponding adequacy criterion is expressed by $M\left\{\left\|\Phi_v^c-\Phi_v^{\exp}\right\|\right\}$, where $M\{\|\cdot\|\}$ denotes the mathematical expectation of the random variable $\|\cdot\|$. For other distribution functions, more complicated methods of estimation are used, for example, the maximum likelihood method.

6.2.2 The Search for the Identified Solutions

The formulation and solution of the identification problem are based on the PSI method. In accordance with the algorithm of the PSI method, we specify the values Φ_v^{**} and find vectors meeting conditions (1.1), (1.2), and (6.1). The vectors α_{id}^i belonging to the feasible solution set D_α will be called *adequate vectors.*

The restoration of the parameters of a specific model on the basis of (1.1), (1.2), and (6.1) is the main purpose and essence of multicriteria parametric identification. In performing this procedure for all structures (mathematical models), we thus carry out multicriteria structural identification.

The vectors α_{id}^i that belong to the set of adequate vectors and have been chosen by using a special decision-making rule will be called *identified vectors.*

The role of the decision-making rule is often played by informal analysis of the set of adequate vectors. If this analysis separates out several equally acceptable vectors α_{id}^i, the solution of the identification problem is not unique.

The identified vectors α_{id}^i form the identification domain $D_{id}=\bigcup_i \alpha_{id}^i$.

By carrying out additional physical experiments, revising constraints Φ_v^{**}, and so forth, one can sometimes reduce the domain D_{id} and even obtain a domain that contains only one vector. Unfortunately, this is far from usual. The nonunique restoration of variables is a compensation for the discrepancy between the physical object and its mathematical model, incompleteness of physical experiments, and so forth.

If a mathematical model is sufficiently good (i.e., it correctly describes the behavior of the physical system), then multicriteria parametric identification leads to a nonempty set D_α. The most important factors that can lead to an empty D_α are imperfection of the mathematical model and lack of information about the domain in which the desired solutions should be searched for.

The search for the set D_α is very important, even in the case where the results are not promising. It enables the researcher to judge the mathematical model objectively (not only intuitively), to analyze its advantages and drawbacks on the basis of all proximity criteria, and to correct the problem formulation. Thus, multicriteria identification includes the determination and informal analysis of the feasible solution set D_α with regard to all basic proximity criteria, as well as finding identified solutions $\boldsymbol{\alpha}^i_{id}$ belonging to this set.

Multicriteria identification is often the only way to evaluate the quality of the mathematical model and hence to optimize this model.

6.2.3 Operational Development of Prototypes

We will discuss the problems of perfecting engineering systems (machines). These problems are mainly related to the operational development of a prototype of a machine designed for mass production. First, the machine is tested. The structure of the test is determined by the type of machine (airplane, automobile, ship, machine tool). For example, automobiles are subjected to laboratory (bench) tests including strength, fatigue, and vibration investigations of both individual units and the automobile as a whole. Much attention is paid to road tests. These are carried out on proving grounds where the automobile is tested on properly profiled road sections in different conditions depending on the load carried and the speed of the automobile. Apart from this, automobiles are tested on ordinary roads under conditions close to operational ones. Thus, automobiles are subjected to bench and road tests. These tests are aimed at detecting imperfections with subsequent operational development of the prototype so as to satisfy the customer's demands. Operational development is aimed at increasing durability and reliability, reducing vibrations and noise, and so forth.

It is very important to make the process of operational development as short as possible. This is the main problem faced by experts of automobiles and other machines. We suggest carrying out the prototype's operational development in two stages. In the first stage, accelerated tests (for instance, bench tests) are performed. These tests make it possible to identify the mathematical model of the object and to determine its parameters.

In the second stage, after multicriteria identification, the expert formulates and solves the multicriteria optimization problem. To do this, one uses the mathematical model whose adequacy was established in the first stage. Based on the optimization results, improvements to the prototype are made, and then the tests are reproduced. This cycle is repeated until the expert decides to terminate the operational development. Thus, in the first stage, the set D_α is found as a result of multicriteria identification. In the second stage, the optimization problem is solved: we construct the parallelepiped Π in D_α, determine the vector of performance criteria, and find the feasible solution set D.

A disadvantage of traditional optimization is the significant discrepancy between the mathematical model and the physical system, as well as improperly specified constraints. Therefore, the results of traditional optimization are very often of no practical value. In our approach, we obtain a confirmed model and the set D_α resulting from multicriteria identification. This, to a sufficient extent, justifies the optimization performed at the second stage and substantiates the recommendations for improving the prototype of a machine. In addition, this approach is expected to significantly reduce the number of expensive and time-consuming tests during the operational development of machines.

6.2.4 Conclusion

The formulation and solution of the multicriteria identification problem combined with informal analysis of the obtained results make it possible to determine:

- The adequacy of the mathematical model (how much it could be trusted);
- The sets of adequate and identified solutions;
- The advantages of one or another model.

6.3 Vector Identification of a Spindle Unit for Metal-Cutting Machines

6.3.1 Introduction

The precision, reliability, and chatter stability of metal-cutting machines depend strongly on the characteristics of their spindle units. This calls on conducting the dynamic and thermal analyses of the spindle units in the design and experimental development stages, using the data obtained by testing a prototype. The dynamic and thermal analyses of spindle units that are carried out in the design stage include the determination of the frequency characteristics and the thermal field, which are used for calculating the dynamic and thermal displacements of the spindle. An accuracy of the calculated frequency and the thermal characteristics of spindle units depends on the reliability of the information about the stiffness and damping characteristics of its supports and the coefficients of convective heat transfer between the surfaces of the spindle and its casing and the environment. An improvement in the reliability of the results in the design stage, obtained by optimizing the structure with the help of mathematical models, can be attained by identifying the model parameters in testing the prototype. In so doing, the identification procedure includes a solution of the direct problems of the dynamics and the heat transfer of the spindle units [1, 6].

6.3.2 Experimental Determination of the Characteristics of a Spindle Unit

Experimental measurements of the dynamic and thermal characteristics of the spindle unit of a multipurpose machine tool were conducted on a special test bed. In the front support of the spindle unit [Figure 6.2(a)], there were mounted three angular-contact ball bearings (#7016AC/P4, d = 80 mm, FAG, Sweden), and the rear support was equipped with a double-row plain roller bearing (#3014, d = 75 mm, FAG) installed in the "triplex" scheme. The supports were lubricated with an NBU15-like lubricant (Kluber, Germany). The spindle casing was rigidly attached to a massive measurement plate isolated from external dynamic disturbances. The spindle was driven by a regulated motor via a belt transmission.

The dynamic characteristics of the spindle unit were determined in the form of amplitude-phase frequency characteristics, an artificial dynamic load being applied to the spindle with the help of a contactless electromagnetic vibrator. From the physical viewpoint, the method is based on the creation of a dynamic load $P(t) = P_0 + P_1 \cdot \cos(\omega t)$, where $P_0 = \text{const}$, P_1 is a variable harmonic component, and ω is a regulated disturbance frequency. The load was applied

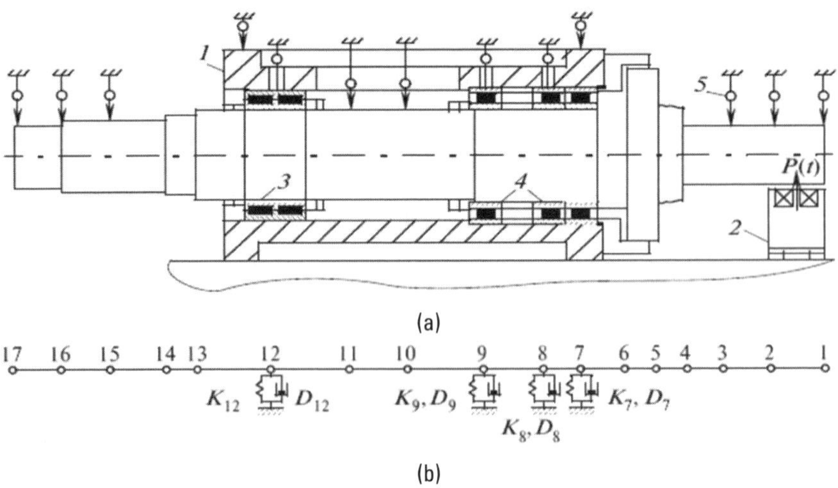

Figure 6.2 Test bed. (a) 1 is the spindle casing, 2 is the electromagnetic vibrator, 3 is the roller bearing, 4 are ball bearings, and 5 is the transducer. (b) Analytical model for analyzing the dynamic model of the spindle; 1 through 17 are nodes.

in the radial direction to the measuring mount 40 mm in diameter, which was inserted in the conical orifice of the spindle. The constant component P_0 = 700N was chosen to ensure operation on the linear portion of the spindle unit static characteristic. The amplitude of the vibrator variable force P_1 = 250N was chosen so as to guarantee a stable display of the spindle vibration amplitude on the vibration analyzer screen (model 2031, B&K, Denmark) with the help of a contactless vortex-current transducer WSG69-5 via an amplifier WSM6983 (Roitlinger, Germany). The measuring system error is 1.0 μm. The studies showed that the use of contactless electromagnetic vibrators guarantees stable results in determining the frequency characteristics. The temperature field of the spindle unit was measured with the help of a digital contact thermal probe (SKF, Sweden).

The temperature was measured to an accuracy of 1°C. Several points of the spindle and its casing were chosen for measuring vibration displacements and the temperature [Figure 6.2(a)]. To measure the temperature of the outer bearing rings, special through orifices were drilled in the zone of the supports.

Figure 6.3 shows the experimental (curve 1, averaged for 10 measurements) and calculated (curve 2) amplitude-phase frequency characteristics for a point on the front of the spindle. The characteristic has a resonance at a frequency of 345 Hz, which corresponds to the first shape of lateral flexural vibrations of the spindle. The resonance at a frequency of 560 Hz is related to the spindle casing vibrations due to its insufficiently stiff attachment to the

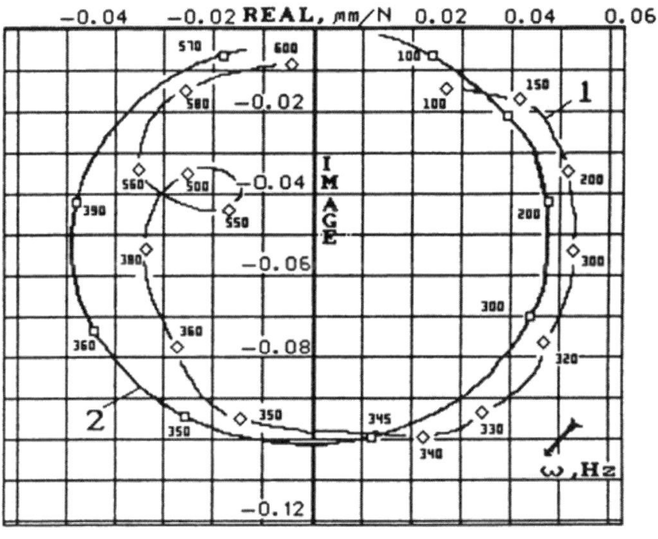

Figure 6.3 Experimental (1) and computed (2) amplitude–phase–frequency characteristics of the spindle at node 1.

mounting plate and was subsequently removed by introducing an additional fixture.

The stationary temperatures of the spindle unit (Figure 6.4) were measured after its 1-hour operation at a rotational frequency of 3,000 1/min (curves 1 and 2 correspond to the experiment and calculation, respectively).

6.3.3 Construction of Mathematical Models

To solve the problem of identification, mathematical models of the dynamic and thermal systems of the spindle unit were developed on the basis of the finite element method. The experience of preceding studies indicates that the bar-structure design models are the most natural and appropriate dynamic and thermal models of spindles with a lined casing. Therefore, the dynamic system of the spindle is presented in Figure 6.2(b) as a linear bar system with distributed mass that rests on concentrated elastoviscous supports. The displacement of each cross-section of the spindle has two components: the radial displacement and the angle of inclination of the spindle axis in the plane of the drawing. Dynamic displacements of the spindle are described by the following system of linear differential equations in the matrix form

$$[M]\cdot\{\ddot{X}\}+[D]\cdot\{\dot{X}\}+[K]\cdot\{X\}=\{P(t)\} \tag{6.2}$$

where $[M]$ is a $2n\times 2n$ matrix of the spindle masses, $[D]$ is a $2n\times 2n$ matrix of damping coefficients of the supports, $[K]$ is a $2n\times 2n$ matrix of stiffness coefficients of the spindle and supports, $\{X(t)\}$ is a 2n vector of the dynamic displace-

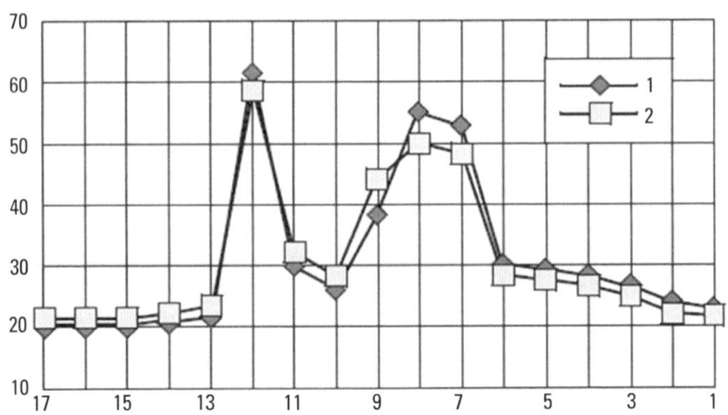

Figure 6.4 Node temperatures of the spindle: experimental (1) and computed (2) data. Vertical axis: Temperature °C; Horizontal axis: Node numbers are on the spindle axis.

ments of the spindle, $\{P(t)\}$ is a 2n vector of the dynamic load generated by the vibrator, and n is the number of nodes in the analytical model of the spindle.

The stationary motion is sought as $\{X(t)\} = \{X_1\}e^{i\omega t}$. Substituting it in (6.2), we find the complex amplitudes of forced vibration

$$\{X(i\omega)\} = \left([K] - \omega^2 \cdot [M] + i\omega \cdot [D]\right)^{-1} \cdot \{P\} \tag{6.3}$$

This expression (6.3) is used for calculating the frequency characteristic of the spindle's dynamic system (curve 2 in Figure 6.3).

In developing the bar thermal model of the spindle unit [Figure 6.5(b)], it was assumed that heat is mostly generated due to friction in the supports [Figure 6.5(a)]. The power of the heat released in the supports is determined using the Palmgren technique [6] and depends on the type of the bearings mounted on the support, the reduced load acting on the support, the rotational frequency, and the lubricant viscosity. Since the heat release in the supports depends on the lubricant viscosity, which, in turn, depends on the temperature, the stationary temperature of the spindle and its supports is found by iteration. Numerical experiments showed that if the environmental temperature is used as the first approximation of the components of the vector of nodal temperatures, then six to eight iterations suffice.

The thermal interaction between the spindle unit and the environment is taken into account by using the coefficient of convective heat transfer between the spindle and casing surfaces and the environment. The temperature field of a lined spindle unit is found by solving the axisymmetric problem of heat transfer under the assumptions: (1) the heat released by a bearing is distributed in equal parts between the races, (2) the rolling elements and the separator are

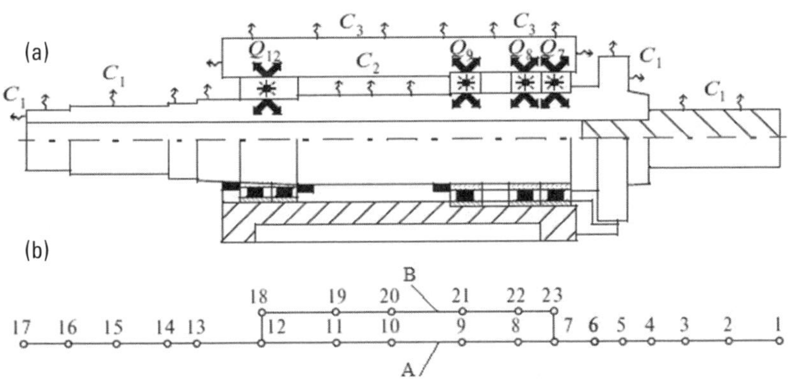

Figure 6.5 Model for the propagation of thermal flows. (a) Q7, Q8, Q9, and Q12 are heat flows in the support, and (b) A is the spindle, B is the spindle casing, and 1 through 23 are the nodes.

viewed as a continuous ring possessing certain thermophysical properties, and (3) the thermal resistance of the joints is ignored. The solution of the stationary problem of heat transfer by the finite element method reduces to that of the system of linear algebraic equations

$$[C]\cdot\{T\}=\{Q\} \tag{6.4}$$

where $[C]$ is an $n\times n$ matrix of the spindle unit heat conductivity, $\{T\}$ and $\{Q\}$ are n-dimensional vectors of the nodal values of the temperature and the thermal load, respectively; and n is the number of nodes in the calculation model.

From system (6.4) we can find the vector of nodal temperatures

$$\{T\}=[C]^{-1}\cdot\{Q\} \tag{6.5}$$

This expression is used for calculating the spindle temperature field (Figure 6.4).

6.3.4 The Identified Parameters of the Models

Since inertial and stiffness parameters of a spindle can be easily expressed through its geometric dimensions, the values of the stiffness coefficients K_7, K_8, K_9, and K_{12} (N/mm) and the damping factors D_7, D_8, D_9, and D_{12} (N·mm/s) of the supports are the identified parameters of the dynamical model [Figure 6.2(b)].

As far as solving the heat transfer problem, the greatest indefiniteness is related to the conditions of the thermal interaction between the spindle unit and the environment, and the identified parameters of the thermal model are the values of the coefficients of convective heat removal from the open spindle surface, C_1, the spindle surface between the front and rear supports, C_2, and the spindle casing surface, C_3 [see Figure 6.5(a)]. Coefficients C_1, C_2, and C_3 are given in Wt/(m$^2\cdot$ °C).

6.3.5 Adequacy Criteria

The set of criteria determining the static and dynamic behavior of the spindle over the range of frequencies from 0 to 600 Hz (criteria $\Phi_1 - \Phi_{13}$) was determined by analyzing the static elastic line of the spindle and its first shape of vibration (345 Hz). In the second problem, it is reasonable to estimate the correspondence between the thermal model and the actual thermal characteristics of the spindle unit according to the temperature values at the structure points (criteria $F_1 - F_{23}$).

The entire list of criteria and their experimental values $\Phi_v^{\exp}, v=\overline{1,13}$ and $F_v^{\exp}, v=\overline{1,23}$ are given in Tables 6.1 and 6.2. The amplitudes of dynamic

Table 6.1
List of Criteria

$\Phi_v^{\exp}$	**Physical Meaning**	**Experimental Value**
$\Phi_1^{\exp}$	The first natural frequency (Hz)	345
$\Phi_2^{\exp}$	Static displacement at node 1 (μm)	50
$\Phi_3^{\exp}$	Dynamic displacement at node 1 (μm)	20
$\Phi_4^{\exp}$	Vibration phase at node 1 (°)	86
$\Phi_5^{\exp}$	Static displacement at node 3 (μm)	27
$\Phi_6^{\exp}$	Dynamic displacement at node 3 (μm)	9
$\Phi_7^{\exp}$	Vibration phase at node 3 (°)	79
$\Phi_8^{\exp}$	Static displacement at node 10 (μm)	–7
$\Phi_9^{\exp}$	Dynamic displacement at node 10 (μm)	5
$\Phi_{10}^{\exp}$	Vibration phase at node 10 (°)	–105
$\Phi_{11}^{\exp}$	Static displacement at node 15 (µm)	2
$\Phi_{12}^{\exp}$	Dynamic displacement at node 15 (μm)	8
$\Phi_{13}^{\exp}$	Vibration phase at node 15 (°)	128

displacements refer to the first resonance frequency of the spindle. The adequacy criteria $\Phi_v, v = \overline{1,13}$ were defined according to the formula

$$\Phi_v^{**} = \left| \frac{\Phi_v^{\exp} - \Phi_v^c\left(\alpha^i\right)}{\Phi_v^{\exp}} \right| \cdot 100\%$$

where $\Phi_v^c(\alpha^i)$ are the computed values of the criteria. In the same way, we determine $F_v^{**}, v = \overline{1,23}$.

Usually, a model is considered to be adequate to the object in the vth criterion, if the relative discrepancy does not exceed 20%.

In solving the problem of identification, the constraints on the stiffness and damping parameters of the spindle supports were formulated, taking an accuracy of their manufacture and assembly into account, and the constraints on the parameters of convective heat transfer were formulated taking the operation conditions of the unit into account.

Table 6.2
List of Criteria

F_v^{exp}	Physical Meaning	Experimental Value
F_1^{exp}	Temperature at node 1 (°C)	22
F_2^{exp}	Temperature at node 2 (°C)	25
F_3^{exp}	Temperature at node 3 (°C)	26
F_4^{exp}	Temperature at node 4 (°C)	28
F_5^{exp}	Temperature at node 5 (°C)	29
F_6^{exp}	Temperature at node 6 (°C)	30
F_7^{exp}	Temperature at node 7 (°C)	54
F_8^{exp}	Temperature at node 8 (°C)	56
F_9^{exp}	Temperature at node 9 (°C)	38
F_{10}^{exp}	Temperature at node 10 (°C)	25
F_{11}^{exp}	Temperature at node 11 (°C)	30
F_{12}^{exp}	Temperature at node 12 (°C)	62
F_{13}^{exp}	Temperature at node 13(°C)	23
F_{14}^{exp}	Temperature at node 14 (°C)	22
F_{15}^{exp}	Temperature at node 15 (°C)	21
F_{16}^{exp}	Temperature at node 16 (°C)	21
F_{17}^{exp}	Temperature at node 17 (°C)	21
F_{18}^{exp}	Temperature at node 18 (°C)	37
F_{19}^{exp}	Temperature at node 19 (°C)	33
F_{20}^{exp}	Temperature at node 20 (°C)	32
F_{21}^{exp}	Temperature at node 21 (°C)	34
F_{22}^{exp}	Temperature at node 22 (°C)	36
F_{23}^{exp}	Temperature at node 23 (°C)	36

6.3.6 Solution of the Identification Problems

We performed 512 trials each in the initial parallelepipeds Π_1^1 and Π_2^1 corresponding to the dynamic and heat models, respectively. The calculations yielded acceptable discrepancies in static (5–10%) and dynamic (9–18%) criteria. However, the discrepancies in thermal criteria turned out to be unacceptably large (34–49%). **Thus, the correctness of the specification of the limiting variation in the convective heat transfer coefficients of the thermal model was doubtful.**

Further modifications of the boundaries of the parallelepiped Π_2^1 for the thermal model were essentially based on an analysis of histograms of feasible solutions and graphs of criterion versus design variable and criterion versus criterion. Multicriteria analysis demonstrated a substantial effect of the heat removal from the surfaces of the spindle unit elements on the temperature field. In view of this, we analyzed and refined the values of the heat exchange coefficients with an allowance for the natural air cooling of the rotating spindle. As a result, we determined the boundaries of a new parallelepiped Π_2^2. After we refined the heat exchange coefficients, the discrepancies with respect to the thermal criteria did not exceed 19%.

For adequacy criteria $\mathbf{\Phi}_v^{**}$ and F_v^{**} not exceeding 20%, we determined the feasible sets D_α for the dynamic and thermal models, which consisted of four and six vectors, respectively. After a informal analysis, the vector $\boldsymbol{\alpha}_{id}^{41} = (K_7; D_7; K_8; D_8; K_9; D_9; K_{12}; D_{12}) = (1.1 \cdot 10^5; 3.5; 8.8 \cdot 10^4; 3.9; 1.5 \cdot 10^5; 4.2; 2.31 \cdot 10^5; 7.8)$ for the dynamic model and the vector $\boldsymbol{\alpha}_{id}^{90} = (C_1; C_2; C_3) = (137; 74; 50)$ for the thermal model were adopted as the best ones. These vectors were chosen for the following reasons:

- The vector $\boldsymbol{\alpha}_{id}^{41}$ corresponds to minimum discrepancies with respect to the static displacement Φ_2 (6%) and the dynamic displacement Φ_3 (11%) of node 1 of the spindle front end. These displacements are known to influence the precision of machining most strongly.
- The vector $\boldsymbol{\alpha}_{id}^{90}$ most accurately reflects the thermal state of the spindle unit near the supports (the temperature discrepancies do not exceed 13%), which is of great practical importance for properly choosing the type of lubricant. The node temperature of the spindel for vector $\boldsymbol{\alpha}_{id}^{90}$ is shown in Figure 6.4, plot 2.

The obtained solutions were analyzed for stability with respect to small parameter variations in the vicinity of the identified vectors. This was done by constructing parallelepipeds centered at $\boldsymbol{\alpha}_{id}^{41}$ and $\boldsymbol{\alpha}_{id}^{90}$ and conducting 128 trials. The stability of the solutions was demonstrated by small variations in the values of the criteria. Thus, having constructed reliable mathematical models, we can pass to the next stage of the experimental development of the spindle, namely, its optimization.

6.3.7 Solution of the Optimization Problem

To determine the feasible sets for the design variables, we analyzed the adequate vectors and constructed parallelepipeds $\overline{\Pi}_1 \subseteq \Pi_1^1$ and $\overline{\Pi}_2 \subseteq \Pi_2^2$ for the dynamic and the thermal models, respectively. The performance criteria to be minimized were Φ_2, Φ_3, F_7, F_8, F_9, and F_{13} (see Tables 6.1 and 6.2). These criteria characterize the static and dynamic stiffness of the spindle unit and the thermal state of its supports. Optimization was accomplished by varying the support stiffness coefficients K_7, K_8, K_9, and K_{12} and the heat exchange coefficients C_1, C_2, and C_3. We performed 256 trials in each of the parallelepipeds $\overline{\Pi}_1$ and $\overline{\Pi}_2$.

The calculations revealed a set consisting of three Pareto optimal vectors of the spindle unit. The expert makes a decision, taking into account the importance of the each criterion. In our case vector α^{114} was preferred, since it is close to the prototype as far as the temperature criteria are concerned and exceeds it in the static and dynamic stiffness by 8% and 12.5%, respectively.

6.3.8 Conclusion

In practice, the stiffness of the bearing was attained by regulating the spindle supports preloading. Later, the results of the experimental development of the spindle unit prototype with the optimal design variables were checked experimentally and proved to be true to an accuracy of measurements.

Thus, in this study we have formulated and solved the problem of vector identification of the spindle unit in 36 static, dynamic, and thermal adequacy criteria. Eleven stiffness, damping, and thermophysical parameters were identified. As a result, the boundaries of the parameter variations in the identification problem were found. The vectors of the dynamic and thermal models, which provide the best possible representation of the static, dynamic, and thermal behavior of the prototype and agree satisfactorily with experimental results, were identified. The solution of the identification problem allowed us to: (1) estimate the quality of the mathematical models of the spindle unit in all major proximity criteria, and (2) formulate and solve the optimal design problem on the basis of static, dynamic, and thermal criteria.

References

[1] Statnikov, R. B., and J. B. Matusov, *Multicriteria Analysis in Engineering,* Dordrecht/Boston/London: Kluwer Academic Publishers, 2002.

[2] Statnikov, R. B., et al., "Multicriteria Analysis Tools in Real-Life Problems," *Journal of Computers & Mathematics with Applications,* Vol. 52, 2006, pp. 1–32.

[3] Statnikov, R. B., *Multicriteria Design: Optimization and Identification,* Dordrecht/Boston/London: Kluwer Academic Publishers, 1999.

[4] Dobrokhodov, V., and R. Statnikov, "Multi-Criteria Identification of a Controllable Descending System," *Proceedings of the First IEEE Symposium on Computational Intelligence in Multi-Criteria Decision-Making (MCDM 2007),* Honolulu, HI, 2007.

[5] Yakimenko, O. A., and R. B. Statnikov, "On Multicriteria Parametric Identification of the Cargo Parafoil Model with the of PSI Method," *Proceedings of the 18th AIAA Aerodynamic Decelerator Systems Technology Conference and Seminary (AIAA 2005),* Munich, Germany, May 23–26, 2005.

[6] Zverev, I. A., "Vector Identification of the Parameters of Spindle Units of Metal-Cutting Machines," *Journal of Machinery Manufacture and Reliability, Russian Academy of Sciences,* No. 6, 1997, pp. 58–63.

[7] Eykhoff, P., *System Identification, Parameter and State Estimation,* New York: Wiley, 1997.

[8] Ljung, L., *System Identification: Theory for the User,* 2nd ed., Upper Saddle River, NJ: Prentice Hall, 1999.

[9] Norton, J., *An Introduction to Identification,* London: Academic International Press, 1986.

[10] Klein, V., and E. A. Morelli, *Aircraft System Identification: Theory and Practice,* AIAA Education Series, Reston, VA: American Institute of Aeronautics and Astronautics, Inc., 2006.

[11] Taylor, L. W., K. W. Iliff, and B. G. Powers, "A Comparison of Newton-Raphson and Other Methods for Determining Stability Derivatives from Flight Data," *AIAA and FTSS Conference,* Houston, TX, 1969, pp. 69–315.

7

Other Multicriteria Problems and Related Issues

In this chapter, we present several additional multicriteria problems from different fields of human activity. The first problem considers the universal question of how to search for the compromise solution when the desired solution is unattainable. The second problem delves into optimal design of controlled systems. The third problem focuses on multicriteria analysis when mathematical models are not available but can be approximated from observational data. The latter problem appeals to disciplines such as biology, economics, materials science, and information science, where the exact mathematical models are unknown. Finally, we consider multicriteria optimization of large-scale systems in parallel mode and discuss issues related to choosing the number of trails in real-life problems.

7.1 Search for the Compromise Solution When the Desired Solution Is Unattainable

This section can be considered as a special case of prototype improvement. An expert a priori defines desired values of criteria vector as $\mathbf{\Phi}^{W} = (\mathbf{\Phi}_1^{W},\ldots,\mathbf{\Phi}_k^{W})$. Let the desired values of local criteria $\mathbf{\Phi}_v^{W}$, $v = 1, \ldots, k$ be *unattainable* (e.g., vector α^{W} does not exist in design variable space [1]). There are two possible approaches in this situation: The first approach is that an expert can make concessions and accept a compromise solution, taking into account the most important criteria. In *parametric optimization* we can vary the object's parameters and change the constraints on them. An example of searching for the compromise solution in parametric optimization problem is given next.

The second approach is to not accept a compromise solution within a given object. In *structural optimization* we can vary the structure of the object. In this case, it is possible that the expert's wishes can be realized using the other structures, which are different from the structure of the prototype. A search for a better structure in the sense of proximity to the given goal Φ^W is the main problem of structure optimization. The multicriteria analysis can answer the important issue of how to improve the prototype and by how much it can be improved in order to get closer to the desired goal.

Finally, we would like to mention that improvement of the prototype depends on: (1) the constructive scheme (topology) of the object, (2) the physical and chemical properties of materials, and (3) the design variable, functional, and criteria constraints. Generally, the approach for prototype improvement is determined by the significance of the object. For example, in order to reduce the mass of the object, to increase its reliability and durability, it might be advisable to use new materials. On the other hand, the cost of the object may increase substantially.

7.1.1 Definition of the Solution That Is the Closest to the Unattainable Solution

Consider the vibratory system within the limits of: (1) the equations (2.3) (Section 2.4), and (2) the given design variable constraints (2.4), functional constraints (2.6), and criteria constraints (Section 3.2, Figure 3.7). Let the desired criteria vector Φ^W be

$$\Phi^W = \left(\Phi_1^W, \Phi_2^W, \Phi_3^W, \Phi_4^W, \Phi_5^W, \Phi_6^W\right) = (34; 29; 3.0; 920; 1.5; 0.86)$$

Obviously, the values of Φ^W cannot be reached: as follows from Table 4.1 (Problem Oscillator, left column) min Φ_4 = (α_3 = 950 + α_4 = 30) = 980 > Φ_4^W = 920. It is necessary to identify the solution that is the closest to Φ^W. As mentioned in Section 3.2.2, we have obtained eight feasible solutions, including four Pareto optimal solutions. Unattainable solution and Pareto optimal solutions are illustrated in Figure 7.1. In this figure vector Φ^W is designated as 0 (blue). The graph fourth criterion versus third criterion is shown in Figure 7.2. The expert was not satisfied with the obtained solutions. However, the analysis of feasible solutions allowed the expert to correct the initial constraints on the design variables (Problem Oscillator 1, right column of Table 4.1).

One hundred ninety-four feasible solutions, including 25 Pareto optimal solutions, were obtained in a new statement (see Sections 4.1.1 and 4.1.2). The fragment of the new table of Pareto optimal solutions is displayed in Figure 7.3.

The analysis of this table allowed the expert to give preference to the compromise solution Φ^{273} = (35.11; 30.196; 3.176; 949.9; 1.76; 8.54), red rectangle. Values of α^{273} = (1.1 · 10^6; 4.53 · 10^4; 9.00 · 10^2; 49.71; 105) are

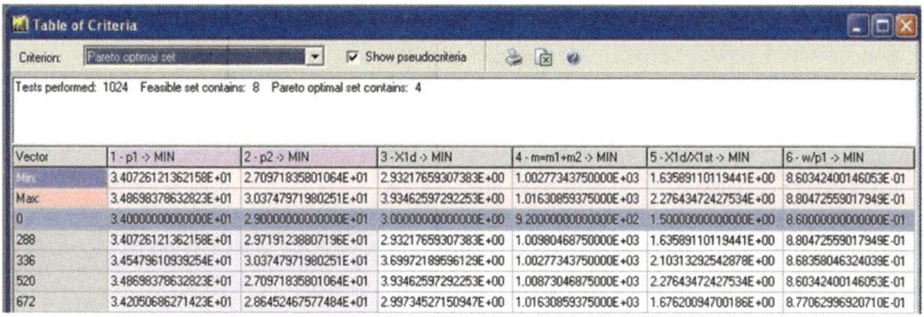

Vector	1 - p1 -> MIN	2 - p2 -> MIN	3 - X1d -> MIN	4 - m=m1+m2 -> MIN	5 - X1d/X1st -> MIN	6 - w/p1 -> MIN
Min	3.40726121362158E+01	2.70971835801064E+01	2.93217659307383E+00	1.00277343750000E+03	1.63589110119441E+00	8.60342400146053E-01
Max	3.48698378632823E+01	3.03747971980251E+01	3.93462597292253E+00	1.01630859375000E+03	2.27643472427534E+00	8.80472559017949E-01
0	3.40000000000000E+01	2.90000000000000E+01	3.00000000000000E+00	9.20000000000000E+02	1.50000000000000E+00	8.60000000000000E-01
288	3.40726121362158E+01	2.97191238807196E+01	2.93217659307383E+00	1.00980468750000E+03	1.63589110119441E+00	8.80472559017949E-01
336	3.45479610939254E+01	3.03747971980251E+01	3.69972189596129E+00	1.00277343750000E+03	2.10313292542878E+00	8.68358046324039E-01
520	3.48698378632823E+01	2.70971835801064E+01	3.93462597292253E+00	1.00873046875000E+03	2.27643472427534E+00	8.60342400146053E-01
672	3.42050686271423E+01	2.86452467577484E+01	2.99734527150947E+00	1.01630859375000E+03	1.67620094700186E+00	8.77062996920710E-01

Figure 7.1 Table of criteria. Unattainable solution is vector #0.

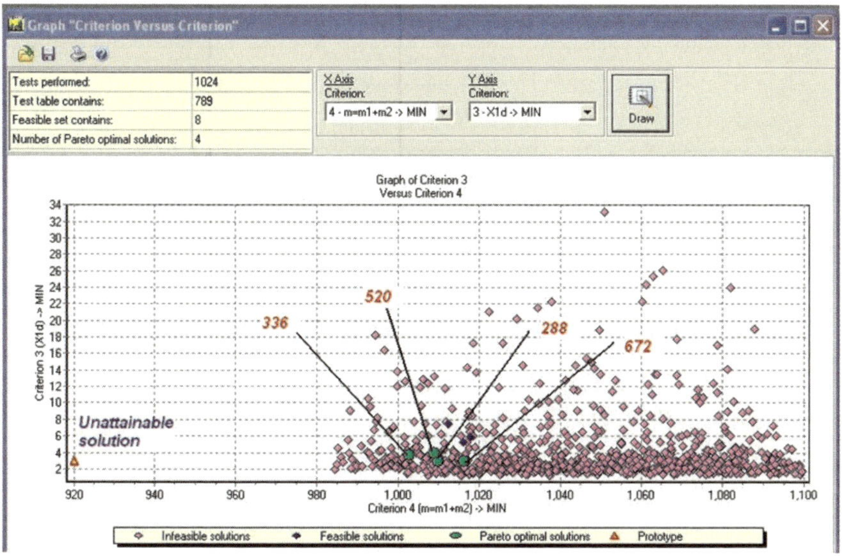

Figure 7.2 Projections of the multidimensional points $\Phi_v(\alpha^j)$, $v = 1, \ldots, 6$; $j = 1, \ldots, 5$ onto the plane fourth criterion versus third criterion. Green and blue points are feasible solutions; green points are Pareto optimal solutions. Unfeasible solutions are magenta.

shown in Figure 7.4 (red rectangle). The unattainable solution, compromise solution 273, and solutions 734, 260, 546 are shown in the graph fourth criterion versus third criterion (see Figure 7.5). The expert also accepted solutions 260, 397, and 734. The values of criteria of these vectors are illustrated in Figure 4.2.

7.1.2 Conclusion

At the expense of the correction of design variable constraints in the initial statement of problem, the compromise solution was found.

Table of Criteria

Criterion: Pareto optimal set ☑ Show pseudocriteria

Tests performed: 1024 Feasible set contains: 194 Pareto optimal set contains: 25

Vector	1 - p1 -> MIN	2 - p2 -> MIN	3 - X1d -> MIN	4 - m=m1+m2 -> MIN	5 - X1d/X1st -> MIN	6 - w/p1 -> MIN
Min	3.30732193449298E+01	2.83405264241461E+01	2.04533620836931E+00	9.48715820312500E+02	1.10803742777651E+00	8.52559746815162E-01
Max	3.51881497010251E+01	3.12382655586961E+01	7.96232123851315E+00	1.01497558593750E+03	4.35711592382895E+00	9.07078312731569E-01
0	3.40000000000000E+01	2.90000000000000E+01	3.00000000000000E+00	9.20000000000000E+02	1.50000000000000E+00	8.60000000000000E-01
117	3.48947303958375E+01	2.96512972424981E+01	2.23274963479056E+00	1.00230468750000E+03	1.28819187913502E+00	8.59728665609023E-01
126	3.48474884360768E+01	2.90896966099589E+01	2.78399441319015E+00	9.56992187500000E+02	1.52793443380162E+00	8.60894180509769E-01
130	3.36573300918093E+01	2.97096519433535E+01	3.31134722371459E+00	9.53964843750000E+02	1.69900569466762E+00	8.91336297863408E-01
242	3.38815139751356E+01	2.92973758876430E+01	2.53846782536874E+00	9.64316406250000E+02	1.32327551287289E+00	8.85438591144890E-01
260	3.30732193449298E+01	3.00145633662976E+01	4.05022997047573E+00	9.49130859375000E+02	2.00098763873406E+00	9.07078312731569E-01
273	3.51144179444971E+01	3.01961114439703E+01	3.17573891780424E+00	9.49912109375000E+02	1.76247307323061E+00	8.54349915394266E-01
313	3.49710101226565E+01	2.90963897171174E+01	2.47350368908885E+00	9.82333984375000E+02	1.40173260818189E+00	8.57853401854241E-01
390	3.43317907791132E+01	2.86573027986895E+01	3.27195027148577E+00	9.53134765625000E+02	1.74109931927011E+00	8.73825667673923E-01
397	3.49747175035373E+01	3.00850833964778E+01	2.93661171263060E+00	9.98447265625000E+02	1.70030965275262E+00	8.57762467901730E-01

Figure 7.3 Table of Pareto optimal solutions. Fragment. Compromise solution is vector #273. Unattainable solution is vector #0.

Design Variable Values

Vectors from: 1 to: 2 Pareto optimal set Show ☐ Hide Constant Variables

Tests performed 1024 Feasible set contains: 194 Pareto optimal set contains: 25

Number of vector	Variable # 1 - k1	Variable # 2 - k2	Variable # 3 - m1	Variable # 4 - m2	Variable # 5 - c
Min	9.88085937500000E+05	4.08835937500000E+04	9.00097656250000E+02	4.34423828125000E+01	8.05078125000000E+01
Max	1.20107421875000E+06	4.95117187500000E+04	9.71191406250000E+02	5.66748046875000E+01	1.18828125000000E+02
117	1.15390625000000E+06	4.80468750000000E+04	9.47656250000000E+02	5.46484375000000E+01	9.71875000000000E+01
126	1.09765625000000E+06	4.49218750000000E+04	9.03906250000000E+02	5.30859375000000E+01	1.04687500000000E+02
130	1.02617187500000E+06	4.24609375000000E+04	9.05859375000000E+02	4.81054687500000E+01	1.12031250000000E+02
242	1.04257812500000E+06	4.81640625000000E+04	9.08203125000000E+02	5.61132812500000E+01	1.00468750000000E+02
260	9.88085937500000E+05	4.12695312500000E+04	9.03320312500000E+02	4.58105468750000E+01	1.18828125000000E+02
273	1.10996093750000E+06	4.53320312500000E+04	9.00195312500000E+02	4.97167968750000E+01	1.05078125000000E+02
313	1.13339843750000E+06	4.70507812500000E+04	9.26757812500000E+02	5.55761718750000E+01	9.44531250000000E+01

Figure 7.4 Table of design variables: Pareto optimal solutions. Fragment. Compromise solution is vector #273.

7.2 Design of Controlled Engineering Systems

An aircraft engine represents a multimode object (i.e., indexes of its efficiency depend on an operating mode). For example, when the plane takes off, the maximal tractive force is necessary, and the engines work in a so-called *maximal* (take-off) mode. Sometimes this mode is named a design mode or a design point. In this mode we have the highest frequency of rotation of the rotors, the highest heat of gas (in front of the turbine), that is, the highest mechanical and thermal loadings. On an interval of rectilinear horizontal flight, the frequency of rotation and the temperature are decreased. As a rule, while designing the

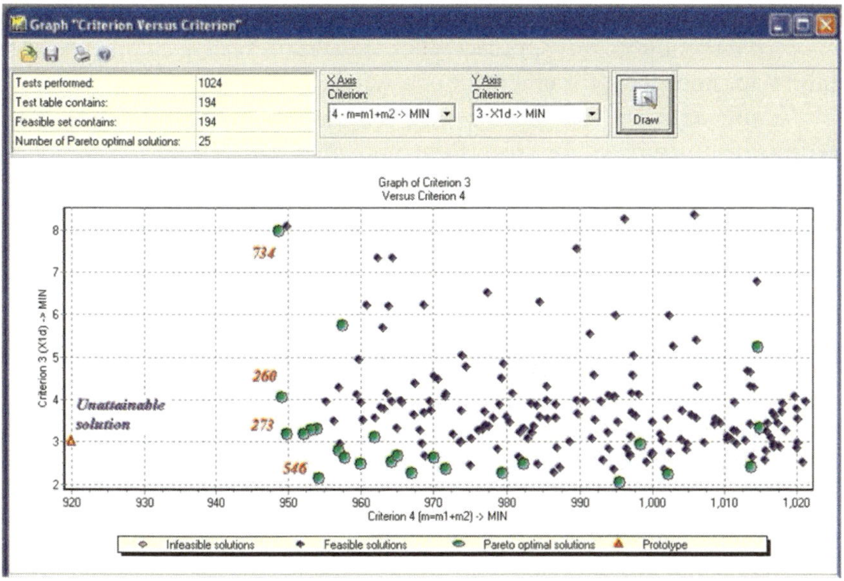

Figure 7.5 Projections of the multidimensional points $\Phi_v(\alpha^j)$, $v = 1, \ldots, 6$; $j = 1, \ldots, 5$ onto the plane fourth criterion versus third criterion. Green and blue points are feasible solutions; green points are Pareto optimal solutions. The compromise solution is vector #273.

engine, only a "maximal" mode is considered. Here the technique of search of optimum decisions is described, taking into account all modes.

Consider an engineering system whose efficiency can be evaluated by a number of particular performance criteria Φ_v, $v = 1, \ldots, k$. The set of criteria Φ_v comprises both "pure design" criteria Φ_{dv}, $v = 1, \ldots, k_1$ (mass, stiffness, safety margins of the structure) and control criteria Φ_{cv}, $v = k_1 + 1, \ldots, k$ (efficiencies of various operating modes). Any particular performance criterion of an engineering system can be represented as a function of the design variable vector α_d and the control variable vector α_c, so that $\Phi_v = \Phi_v(\alpha_d, \alpha_c)$.

It is important to emphasize an essential difference between design variables and control variables. Design variables mostly characterize the sizes of separate parts of a system. These variables cannot be changed purposefully in the course of operating the system (changes in the geometric characteristics caused by mechanical wear or as a result of repair are not considered). The control variables in modern engineering systems can be collected in databases stored in the computer memory. Therefore, in principle, a number of control laws can be implemented, each of which can be chosen depending on the specific task fulfilled by the engineering system.

The traditional approach to the optimization of regulated engineering systems yields a single design variable vector α_d (the design of the system) and the corresponding single control variable vector α_c (the set of control laws). However, rather often, this approach does not allow revealing all potential possibilities for improving the efficiency of a system with controlled variables.

A more efficient approach [2–6] to the optimization of controlled engineering systems employs, for a fixed design variable vector α_d, not a single control variable vector, but a set of vectors, each of which determines the optimal set of control laws for each purpose (operating mode). All the control variable vectors are stored in the airborne computer's memory and may be chosen in accordance with a concrete control purpose, thus implementing the optimal control. When using this approach, one has, first of all, to construct a set $\tilde{D}$ of feasible solutions $\alpha = (\alpha_d, \alpha_{ci}) \in \tilde{D}, i = \overline{1, p_\alpha}$, where p_α sets of control laws (specified by the control variable vectors α_{ci}) correspond to each design (specified by the design variable vector α_d). Then it is necessary to determine a set of Pareto optimal designs $\tilde{P} \subseteq \tilde{D}$ and to select from this set a design $\alpha^0 = (\alpha_d^0, \alpha_{ci}^0) \in \tilde{P}, i = \overline{1, p_{\alpha^0}}$ that is most preferable from the viewpoint of the expert. However, for problems of high dimensionality, in which the number of design variables and control variables can amount to many dozens, it is extremely difficult to construct the feasible set $\tilde{D}$. Therefore, the following algorithm is proposed for solving practical problems.

Stage 1

The determination of the feasible set D consists of design and control vectors, $\alpha = (\alpha_d, \alpha_c)$. As a result, only one set of control laws (control variable vector α_c) corresponds to each feasible design α_d.

Stage 2

To estimate the ultimate possibilities of the system, one must solve the multicriteria problem of optimizing the control variables with respect to the control criteria $\Phi_{cv}, v = \overline{k_1 + 1, k}$ for all feasible designs. In other words, for each fixed α_d from the set D, by varying only control variables α_c, we construct the vectors $(\alpha_d, \alpha_{ci}) \in \tilde{D}$ where p_α Pareto optimal control laws correspond to each α_d. To complete this stage, we determine the set $\tilde{P} \subseteq \tilde{D}$ of the Pareto optimal solutions.

Stage 3

The results of an analysis of the set $\tilde{P}$ are used for choosing the most preferable solution $\alpha^0 = (\alpha_d^0, \alpha_{ci}^0), i = \overline{1, p_{\alpha^0}}$.

If the number of control or/and design variables is large, the construction of the set $\tilde{D}$ requires a rather extensive numerical experiment. Conducting such an experiment is sometimes either complicated or impossible. In this case, in Stage 1, we select from the set $P \subseteq D$ a subset of the most acceptable vectors $\alpha^j = (\alpha_d^j, \alpha_c^j)$. Then for each of the selected α_d^j, we solve the multicriteria control problem in accordance with Stage 2.

In this approach we construct the feasible and Pareto optimal sets taking into account all parameters (design and control variables) and all criteria ("pure design" and control criteria).

This approach that was used for solving the problem of optimal gas dynamic design of a four-stage axial flow compressor for an aircraft gas turbine engine is described next.

7.2.1 Multistage Axial Flow Compressor for an Aircraft Engine

7.2.2.1 Characteristics of the Object of Investigation

A multistage axial flow compressor is intended for increasing the gas pressure in aircraft or ground power plants [4–6]. A schematic diagram of the compressor passage is shown in Figure 7.6. Each stage of the compressor consists of two blade rows: the rotor blade (RB) row and the stator blade (SB) row. Each blade has a complex three-dimensional shape. The rotor blade is shown in Figure 7.7. The shape of the blades is specified by a number of design variables and cannot be purposefully changed in the course of operation. However, the stator blades of the first three stages of this compressor can be rotated about their axes, and, hence, one can control the angles of rotation depending on the operating mode in order to increase the compressor's efficiency. The operating mode of the compressor is characterized by the reduced rotation rate $n_r = n \cdot \sqrt{288/T_{in}}$, where n is the ratio of the current rotation rate to the rotation rate at the maximum power mode and T_{in} is the inlet air temperature.

The gas dynamic efficiency of the compressor depends on the air flow rate $G(n_r)$, the total pressure ratio $\pi(n_r)$, the efficiency $\eta(n_r)$, and the gas dynamic stability margin $\Delta K_s(n_r)$. All these characteristics depend on the operating mode. The values $G(n_r)$ and $\pi(n_r)$ are fixed; the function $\eta(n_r)$ is to be maximized, and the value of the $\Delta K_s(n_r)$ is bounded below, where $\Delta K_s(n_r) \geq \Delta K_{s\,min}$.

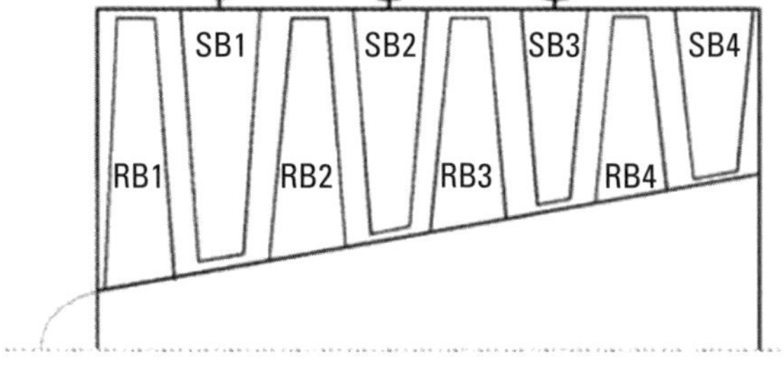

Figure 7.6 Schematic diagram of the compressor passage.

Figure 7.7 General view of the rotor blades. (Dashed line shows a possible change in the blade shape due to a change in design variables.)

We seek the laws of control for the angles of rotation of the first three stator blade rows in the form of linear functions of n_r so that $\alpha_c = (\alpha_{c1},\ldots,\alpha_{c6})$. We consider as the design variables the inlet and outlet angles of all rotor blades and of the first three stator blades in three radial sections; hence, $\alpha_d = (\alpha_{d1},\ldots,\alpha_{d42})$. We have normalized the nominal values and ranges of the parameters to be varied so that the values corresponding to the prototype are equal to 1.5 and the boundaries of the ranges are specified by $\alpha_j^* = 1.0$ and $\alpha_j^{**} = 2.0, j = \overline{1,48}$.

Within the framework of this study, the problem of optimal design of a regulated compressor is viewed as a problem of improving the prototype.

We consider as the performance criteria the difference $\Phi_v(\alpha) = |\eta_v(\alpha) - \eta_v^p|, v = \overline{1,4}$ between the efficiency η_v of the compressor to be designed and the corresponding efficiency η_v^p of the prototype for four operating modes with n_r = 1.0; 0.9; 0.8; 0.7. It is desirable to increase the criteria Φ_v by optimizing the design variables and control variables of the compressor.

The functional constraints were reduced to ensuring: (1) a given air flow rate, (2) a given increase in the pressure to an accuracy of 1%, and (3) the gas dynamic stability margin. In the case under consideration, no criteria constraints were imposed. Design variable and functional constraints form the feasible solution set D.

The choice of the mathematical model of the object under consideration is one of the most important stages in optimizing complex engineering systems. After all, the usefulness of any optimization depends greatly on the adequacy of the mathematical model employed. This study used the two-dimensional axisymmetric model considered in [4]. Since it was identified using the results of an experimental study of the prototype compressor, the calculation of the main gas dynamic characteristics to an accuracy of no less than 0.5% (over the considered range of operating modes) is guaranteed.

The solution of this problem is described in detail in [2, 3, 5, 6]. In Table 7.1 we present the values of the performance criteria in various operating modes for four Pareto optimal control laws obtained for one of the best designs. From Table 7.1 it follows that for the first mode the best value is $\Phi_1 = 0.048\%$ (Pareto optimal control law 1), for the second mode $\Phi_2 = 0.922\%$ (Pareto optimal control law 4), for the third mode $\Phi_3 = 1.34\%$ (Pareto optimal control law 3), and for the fourth mode $\Phi_4 = 2.212\%$ (Pareto optimal control law 4).

This table shows the limiting values of the criteria that can be provided by implementing several control laws obtained from the Pareto optimal set. In fact, in this case, each operating mode is most effective for a specific control law.

An analysis of the obtained results indicates that, first, there exists a set of alternative control laws for each chosen design, and, second, the design variables determine the efficiency (that is captured by four criteria for different modes) of a compressor regulated in various operating modes. If one would work with only the first law, the efficiency will be 0.048%, 0.54%, 0.908%, and 1.315% for the first, second, third, and fourth modes, respectively. In our case, we have the following efficiencies (in order from the first to the fourth modes): 0.048%, 0.922%, 1.34%, and 2.212%. This confirms the validity of the chosen strategy for optimizing a regulated multistage axial flow compressor.

Multicriteria control allows a considerable increase in the compressor efficiency over the entire range of operating modes.

7.2.2 Conclusion

The studies showed that the optimal design of regulated engineering systems should be carried out within the framework of a single optimization investigation strategy, which includes simultaneous multicriteria optimization of the design variables and control laws.

This approach makes it possible to realize the limiting possibilities for improving the efficiency of complex engineering systems by choosing the most preferable designs from the obtained set and realizing (on an on-board

Table 7.1
Four Pareto Optimal Control Laws

Optimal solutions	Φ_1 (%)	Φ_2 (%)	Φ_3 (%)	Φ_4 (%)
1	<u>0.048</u>	0.54	0.908	1.315
2	–0.001	0.872	0.789	0.711
3	–0.006	0.628	1.224	<u>2.212</u>
4	–0.217	<u>0.922</u>	<u>1.34</u>	1.292
Optimal control	0.048	0.922	1.34	2.212

Note: The best value for each mode is underlined.

computer) a set of the families of control laws that are optimal for various purposes and operating modes of the engineering system.

7.3 Multicriteria Analysis from Observational Data

For this class of problems, there are no a priori specified mathematical models [7]. However, there are available observations in the form of tables that give an indication of the behavior of the system under investigation. These problems are often encountered in medicine, biology, economics, materials science, information science, and other fields. An approximate mathematical model is constructed on the basis of the observations with the use of classification and regression algorithms. Some algorithms for constructing approximate criteria functions include regression by neural networks, Support Vector Machine (SVM) regression, and multiple linear regression [8–11]. Next we briefly describe a general strategy for multicriteria analysis from observational data:

- *Step 1: Obtaining observational data and constructing an approximate model.* Suppose we have an experiment with N observations represented by an $N \times M$ matrix, where M is the total number of observed variables (criteria and design variables). The approximate criteria functions are constructed using machine learning algorithms. The approximate criteria functions are evaluated by statistical metrics, such as R^2 (fraction of variance explained by a model) and absolute, relative, and squared errors. Then functions with the best evaluation performance are chosen for the approximate mathematical model.
- *Step 2: Multicriteria analysis: construction of a pseudo-feasible solution set and the search for the best solutions.*

7.3.1 Example

Here we illustrate the process of constructing the approximate feasible solution set in problems in which only observational data are present. We will consider an oscillatory system and imagine that we do not have access to its mathematical model.

The collection of observational data depends on the specifics of the problem being investigated. In the present case, in order to obtain observational data, we referred to the true model[1] and, using the PSI method, conducted 4,000 trials with a random number generator. As a result, we obtained a 4,000

1. The true model is presented by the equations (2.3), design variable constraints (2.4), rigid functional constraints (2.6), and criteria constraints (see Figure 3.7).

× 11 ($M = 6$ criteria and 5 design variables) matrix of observations. See Figures 7.8 and 7.9 for fragments of this matrix.

Using the observational data, we constructed approximate criteria functions by means of machine learning algorithms. In our case, criteria Φ_3 and Φ_5 were determined using generalized neural networks for regression [8], while the remaining four criteria were reconstructed using the SVMTorch algorithm [9–11]. This choice was based on statistical estimates of the criteria functions obtained; the estimates of the best approximate functions are given in Table 7.2. These criteria functions constitute the approximate mathematical model.

Vector	1 - p1	2 - p2	3 - X1d	4 - m=m1+m2	5 - X1d/X1st	6 - w/p1
Min	3.36893057980329E+01	2.73861278752583E+01	1.39795047470178E+00	1.00375000000000E+03	1.31931576049980E+00	6.77451398515785E-01
Max	4.42836195566596E+01	3.79777262656375E+01	2.43905281949274E+01	1.09625000000000E+03	1.41007741126924E+01	8.90490299202059E-01
1	3.93700393700591E+01	3.00000000000000E+01	2.20641177157358E+00	1.05000000000000E+03	1.70996912296952E+00	7.62000762001143E-01
2	3.68642694078475E+01	2.81365716935569E+01	2.52459939526769E+00	1.03500000000000E+03	1.67254709936485E+00	8.13796135984557E-01
3	4.16137875838426E+01	3.25960120260132E+01	3.01714085618395E+00	1.06500000000000E+03	2.67771250986326E+00	7.20914911663752E-01
5	4.10310163632034E+01	3.02765035409749E+01	2.45604621874458E+00	1.03250000000000E+03	2.04158841933143E+00	7.31154201359340E-01
6	3.76796110173626E+01	3.53553390593274E+01	4.75924877522764E+00	1.04750000000000E+03	3.42071005719486E+00	7.96186563236445E-01
7	4.42836195566596E+01	2.97718598069324E+01	1.39795047470178E+00	1.01750000000000E+03	1.31931576049980E+00	6.77451398515785E-01
8	3.36893057980329E+01	3.40846696607486E+01	2.43905281949274E+01	1.06125000000000E+03	1.41007741126924E+01	8.90490299202059E-01
10	3.63779231597019E+01	2.82421501946833E+01	2.71723505246155E+00	1.09625000000000E+03	1.87659045810626E+00	8.24675995611341E-01
11	4.29274655156519E+01	3.79777262656375E+01	2.32288016672301E+00	1.02625000000000E+03	2.12688715265576E+00	6.98853278189965E-01
12	3.59582206257936E+01	2.73861278752583E+01	3.24031835279167E+00	1.03875000000000E+03	2.05557695505222E+00	8.34301572155113E-01
13	4.08248290463863E+01	3.58236421003411E+01	2.94097626076868E+00	1.06875000000000E+03	2.52740147409808E+00	7.34846922834953E-01
14	3.95233032939806E+01	3.09923589903943E+01	4.44071479107186E+00	1.00375000000000E+03	3.31665885958179E+00	7.59045866608246E-01
16	3.37461295829559E+01	3.41958307211519E+01	2.02609066999235E+01	1.02937500000000E+03	1.14284176854256E+01	8.88990837490058E-01

Figure 7.8 Fragment of the matrix corresponding to criteria values for first 16 observations (out of 4,000 total).

Number of vector	Variable # 1 - k1	Variable # 2 - k2	Variable # 3 - m1	Variable # 4 - m2	Variable # 5 - c
Min	1.12812500000000E+06	4.06250000000000E+04	9.56250000000000E+02	3.25000000000000E+01	8.25000000000000E+01
Max	1.94375000000000E+06	4.93750000000000E+04	1.04375000000000E+03	6.75000000000000E+01	1.17500000000000E+02
1	1.55000000000000E+06	4.50000000000000E+04	1.00000000000000E+03	5.00000000000000E+01	1.00000000000000E+02
2	1.32500000000000E+06	4.75000000000000E+04	9.75000000000000E+02	6.00000000000000E+01	9.00000000000000E+01
3	1.77500000000000E+06	4.25000000000000E+04	1.02500000000000E+03	4.00000000000000E+01	1.10000000000000E+02
4	1.21250000000000E+06	4.62500000000000E+04	1.03750000000000E+03	6.50000000000000E+01	1.05000000000000E+02
5	1.66250000000000E+06	4.12500000000000E+04	9.87500000000000E+02	4.50000000000000E+01	8.50000000000000E+01
6	1.43750000000000E+06	4.37500000000000E+04	1.01250000000000E+03	3.50000000000000E+01	1.15000000000000E+02
7	1.88750000000000E+06	4.87500000000000E+04	9.62500000000000E+02	5.50000000000000E+01	9.50000000000000E+01
8	1.15625000000000E+06	4.93750000000000E+04	1.01875000000000E+03	4.25000000000000E+01	8.75000000000000E+01
9	1.60625000000000E+06	4.43750000000000E+04	9.68750000000000E+02	6.25000000000000E+01	1.07500000000000E+02
10	1.38125000000000E+06	4.18750000000000E+04	1.04375000000000E+03	5.25000000000000E+01	9.75000000000000E+01
11	1.83125000000000E+06	4.68750000000000E+04	9.93750000000000E+02	3.25000000000000E+01	1.17500000000000E+02
12	1.26875000000000E+06	4.31250000000000E+04	9.81250000000000E+02	5.75000000000000E+01	1.02500000000000E+02
13	1.71875000000000E+06	4.81250000000000E+04	1.03125000000000E+03	3.75000000000000E+01	8.25000000000000E+01
14	1.49375000000000E+06	4.56250000000000E+04	9.56250000000000E+02	4.75000000000000E+01	1.12500000000000E+02
15	1.94375000000000E+06	4.06250000000000E+04	1.00625000000000E+03	6.75000000000000E+01	9.25000000000000E+01
16	1.12812500000000E+06	4.53125000000000E+04	9.90625000000000E+02	3.87500000000000E+01	9.87500000000000E+01

Figure 7.9 Fragment of the matrix corresponding to design variable values for first 16 observations (out of 4,000 total).

Table 7.2
Statistical Estimates of the Approximate Criteria Functions

Criterion	R^2	Mean Absolute Error	Mean Relative Error	Mean Squared Error
Φ_1	0.999997	0.0361611	3.44387e-005	0.00313826
Φ_2	0.999975	0.0121383	0.000329313	0.000357047
Φ_3	0.810006	0.603323	0.119773	2.35193
Φ_4	0.999997	0.0361611	3.44387e-005	0.00313826
Φ_5	0.797901	0.368809	0.109251	0.778614
Φ_6	0.998594	0.0024371	0.00319643	7.67688e-006

At this stage, we have an approximate model. We employed the PSI method to conduct 1,024 trials using LP_τ sequences. We constructed test tables and obtained eight approximate feasible solutions $\mathbf{\Phi}(\alpha^{i_p})$ that satisfied all constraints. These were vectors #288, #336, #520, #544, #560, #672, #896, and #1008. Since in this example we knew the true model, the vectors α^{i_p} were checked for feasibility by the direct application of the true model and the calculation of the values $\mathbf{\Phi}(\alpha^{i_p})$. Seven of these eight approximate feasible solutions were found to be feasible.[2] The eighth approximate feasible solution #1008 was nonfeasible because of errors in the approximate model.

After constructing and analyzing the approximate feasible solution set, we corrected the design variable constraints and determined a new approximate feasible set. After 1,024 trials with LP_τ sequences, 311 approximate feasible solutions were identified, 218 of which turned out to be feasible. We note that in the same experiment with the true model, there were 258 feasible solutions [7].

In order to analyze the efficiency of the employed approximate model, we can use a metric equal to the number of feasible solutions found with the approximate model over the number of feasible solutions obtained with the true model. In the cases described above, this metric is 7/8 = 0.875 and 218/258 = 0.85, respectively. Further improvement of the efficiency of an approximate model is possible by improving the approximate criteria, especially Φ_3 and Φ_5. However, it is worth noting that we have already approximated the true model fairly well, such that we have preserved the dependences of criteria on design variables and between criteria (see Figures 7.10 through 7.13).

To summarize, multicriteria analysis can be carried out in problems from observational data by constructing an approximate model. This analysis

2. Recall that eight feasible solutions #288, #336, #520, #544, #560, #672, #896, and #968 were found from the true model (see Figure 3.7).

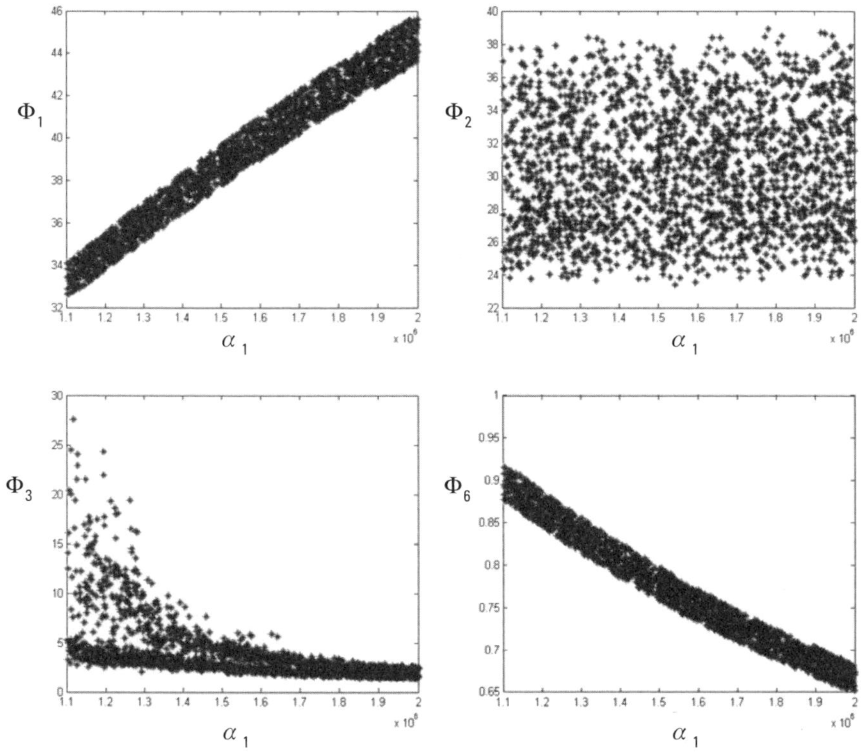

Figure 7.10 The dependences of criteria on the first design variable for the approximate mathematical model. See Figure 7.11 for the true model.

can be used to predict the best solutions and approaches to their subsequent improvement.

7.4 Multicriteria Optimization of Large-Scale Systems in Parallel Mode

7.4.1 Computationally Expensive Problems

For many applied optimization problems, it is necessary to carry out a large-scale numerical experiment in order to construct the feasible set. For this reason, a search for optimal solutions is often not carried out at all. We will mention a few types of difficult problems.

- *The first type:* Problems with a small area of the feasible solution set (when a ratio of the volume of the feasible solution set to the volume of

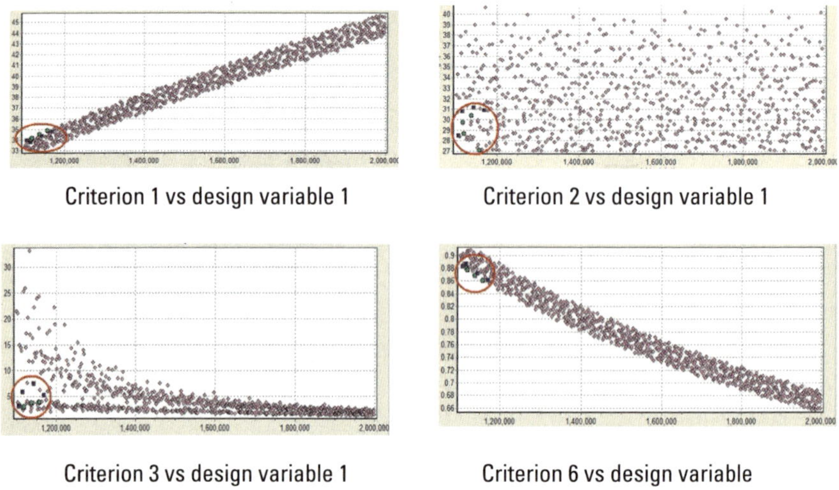

Criterion 1 vs design variable 1

Criterion 2 vs design variable 1

Criterion 3 vs design variable 1

Criterion 6 vs design variable

Figure 7.11 The dependences of criteria on the first design variable for the true mathematical model. The area of feasible solutions (blue and green points) including Pareto optimal solutions (green points) is circled.

the parallelepiped is a small value, e.g., << 0.0001). Even if the time for calculation of one criteria vector is fairly short, it takes a long time to find at least one feasible solution because of the need to carry out a large number of trials.

- *The second type:* Problems with a high dimensionality of the design variable vectors (e.g., thousands of design variables). It is obvious that these problems also require a large-scale numerical experiment with hundreds of thousands or millions of trials.
- *The third type:* Problems with complex mathematical models, where calculating one criteria vector requires a lot of computer time. For example, this includes many problems with finite element models.

The software package MOVI allows us to tackle computationally expensive problems in *parallel mode,* so that the desired number of trials *N* is distributed among *k* computers [1, 7]. Thus, each computer finds a feasible solution set for its own subproblem (by conducting ~*N*/*k* trials). Next all feasible solution sets are combined and a single Pareto optimal solution set is constructed.

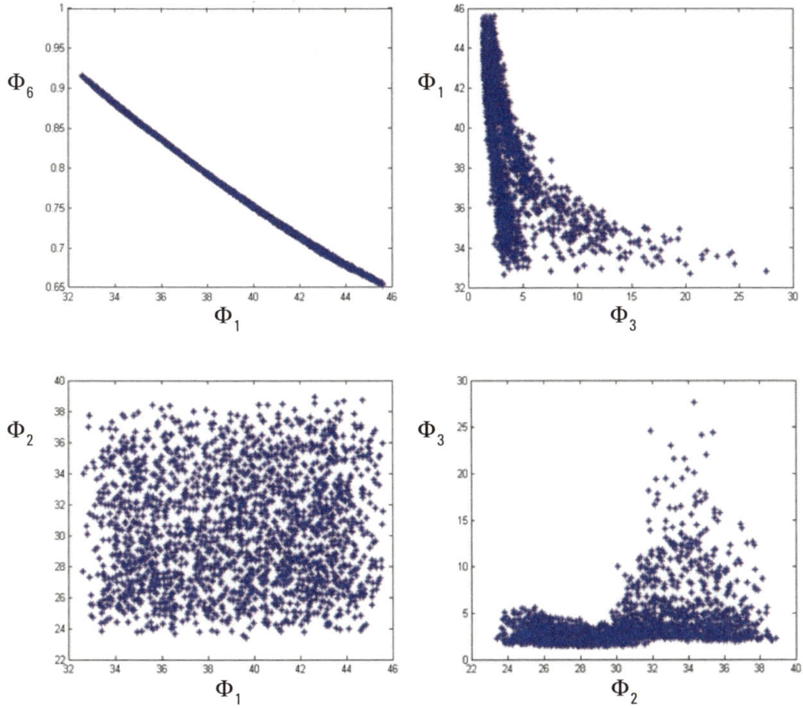

Figure 7.12 The dependences between criteria for the approximate mathematical model. See Figure 7.13 for the true model.

7.4.2 First Example

Consider a system with 1,000 design variables [7]. The design variable vector is given by $\alpha = (\alpha_1, ..., \alpha_{1000})$, $1 \le \alpha_i \le 2$, $i = 1, \ldots, 1{,}000$. We are seeking to minimize simultaneously the following performance criteria $\Phi_v(\alpha)$:

$$\Phi_1 = \sum_1^{1000} \alpha_i,$$

$$\Phi_2 = \sum_{300}^{1000} \alpha_i^2 - \sum_1^{299} \alpha_i^2,$$

$$\Phi_3 = \left(1400 \Big/ \sum_{300}^{1000} \alpha_i^2\right) - \cos\left(\sum_1^{299} \alpha_i\right)$$

$$\Phi_4 = \sum_1^{700} \frac{\alpha_i}{i} - \left(\sin\left(\sum_{701}^{1000} \alpha_i^2\right)\right)^5$$

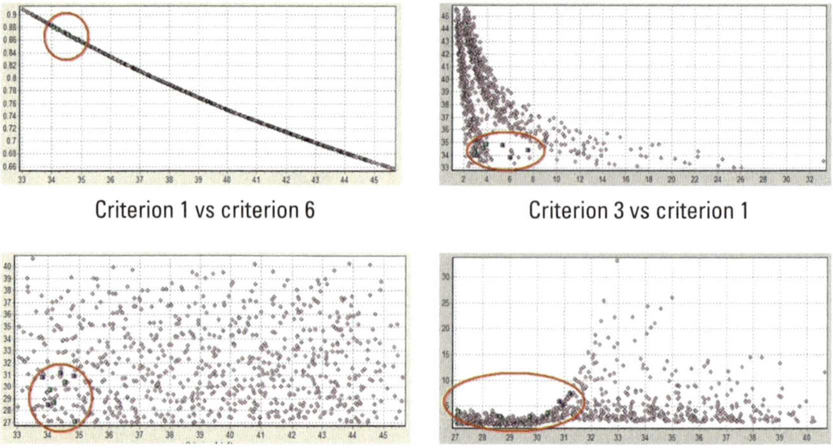

Criterion 1 vs criterion 6

Criterion 3 vs criterion 1

Criterion 1 vs criterion 2

Criterion 2 vs criterion 3

Figure 7.13 The dependences between criteria for the true mathematical model. The area of feasible solutions (blue and green points) including Pareto optimal solutions (green points) is circled.

While analyzing the test tables, we formulated the following criteria constraints:

$$\Phi_1 < 1502.2254,$$
$$\Phi_2 < 930.4528,$$
$$\Phi_3 < 0.162,$$
$$\Phi_4 < 10.3851$$

We investigated the design variable space and criteria space using four computers simultaneously (see Figure 7.14). Each computer conducted 50,000 trials using a random number generator. Four computers conducted a total of 200,000 trials, which resulted in 4,297 feasible solutions. The CPU time was approximately 8 hours per computer using an Intel Xeon 2.4 GHz with 2 GB of RAM. To solve this problem on one computer, 34 hours would be required.

After we combined all 4,297 feasible solutions, we obtained 326 Pareto optimal ones. The efficiency coefficients for the Pareto optimal and feasible solutions are equal to γ_p = 326/200,000 = 0.0016 and γ_f = 4,297/200,000 = 0.021, respectively.

7.4.3 Second Example

The following performance criteria need to be minimized [7]:

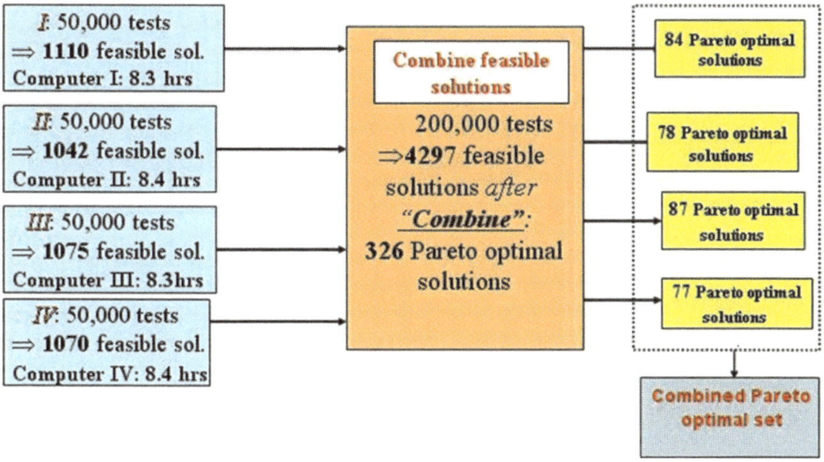

Figure 7.14 Construction of combined feasible and Pareto optimal sets in a parallel mode.

$$\Phi_1 = \sum_{1}^{50} \alpha_i,$$

$$\Phi_2 = \sum_{21}^{50} \alpha_i^2 - \sum_{1}^{20} \alpha_i^2,$$

$$\Phi_3 = \left(1400 \Big/ \sum_{21}^{50} \alpha_i^2\right),$$

$$\Phi_4 = \sum_{15}^{25} \frac{\alpha_i}{i} - \left(\sum_{26}^{50} \alpha_i^2\right)^3$$

We have 50 design variables with the following intervals of variation: $1 \le \alpha_i \le 2$, $i = 1, \ldots, 50$.

We are also given a priori criteria constraints:

$$\Phi_1^{**} = 69.804,$$

$$\Phi_2^{**} = 20.384,$$

$$\Phi_3^{**} = 23.570,$$

$$\Phi_4^{**} = -120{,}600$$

A total of 250,000 trials were conducted on five computers (50,000 trials each) using a random number generator. The combined feasible solution set and Pareto optimal set were constructed. On the first computer, we performed trials with numbers from 1 to 50,000, on the second computer trials with numbers from 50,001 to 100,000, on the third computer trials with numbers from

100,001 to 150,000, and so forth. In general, when we use uniformly distributed sequences (e.g., LP_τ), the values of design variables are guaranteed to be different for each test, while the criteria values may coincide. In this case, MOVI software will mark that these criteria vectors have been obtained, for example, on computers X and Y.

The results of the investigation are presented in Table 7.3.

The combined feasible set contains 87 solutions, and the combined Pareto optimal set contains 57 solutions. For example, data from the first computer are given in the first row: 23 feasible solutions, 20 of which are Pareto optimal solutions; the first computer contributes 14 vectors to the combined Pareto optimal solution set. The contribution of each computer to the combined Pareto optimal solution set is shown in the last column. The coefficients γ for the Pareto optimal and feasible solutions are equal to $\gamma_p = 57/250000 = 0.000228$ and $\gamma_f = 87/250.000 = 0.000348$, respectively. The dependences between criteria obtained on the first computer after carrying out 50,000 trials are shown in Figure 7.15. This analysis shows the complex relationships between the criteria and the localization of the feasible solutions.

7.4.4 Conclusion

In many cases we have difficult optimization problems with stringent constraints, high dimensionality of design variables, and models that require a lot of time to calculate a criteria vector. The possibility of using the PSI method and the MOVI to carry out large-scale numerical experiment in parallel mode in order to construct the feasible set allows to solve many of the problems that until recently were considered impossible to optimize.

Table 7.3

Pareto Optimal Solutions Obtained on Five Computers

Number of a Computer	Feasible and Pareto Optimal Solutions	The Contribution of Each Computer to the Combined Pareto Optimal Solution Set
1	23 (20)	14
2	14 (11)	9
3	13 (10)	8
4	19 (19)	14
5	18 (13)	12

Note: The combined Pareto optimal solution set contains fewer solutions (57) than the sum of Pareto optimal solutions obtained on each computer (73) because 16 solutions were surpassed by solutions from other computers and thus lost their Pareto optimality.

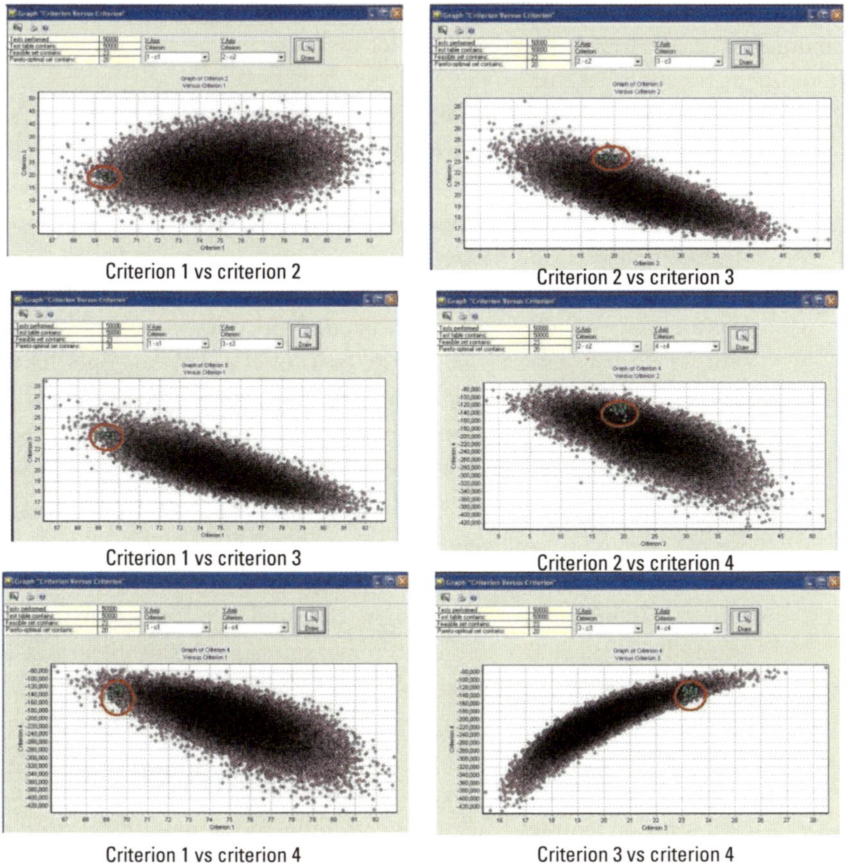

Figure 7.15 Dependences between criteria obtained on the first computer. The regions of the feasible and Pareto optimal solutions are circled. Pareto optimal solutions are green. Unfeasible solutions are magenta.

7.5 On the Number of Trails in the Real-Life Problems

As known, the number of tests depends on constraints, dimensionality of design variable vectors, and the time of calculating one criteria vector for given design variable values. As already noted, when using the PSI method, stating and solving problems is a *single process*. When we mention the number of tests, we mean this particularly complicated process. Indeed, in order to obtain a feasible solution and Pareto optimal sets, we should repeatedly correct initial statements of problems on the basis of the PSI method. The necessity of correcting the statement of the problem is one of the important features of real-life problems. All the problems described in our book prove this fact. We will point to three kinds of problems depending on the time of calculating one criteria vector and the number of performed tests [12].

1. *Problem of a naval ship design* (Section 5.2). Among the particular features of this problem are the high dimensionality of the design variable vector (45 design variables) and the difficulties in improving a reasonably good prototype under strong constraints on six performance criteria, nine pseudo-criteria, and seven functional dependences. Since calculation of one criterion vector took less than 1 second, hundreds of thousands of trails were conducted. Notice that each subsequent experiment was carried out on the basis of the previous one. In the process of solving these problems, experts repeatedly corrected their statement. A similar large-scale experiment was also discussed in Section 5.1. In the latter case, we corrected the problem statement twice.
2. *Problem of a design of a car for shock protection* [13]. Unlike the previous problem where calculation of a criteria vector took less than 1 second, the criteria in this problem are based on the finite element model with thousands of elements and nodes, and it takes approximately several hours to compute one vector of criteria. This problem has 10 criteria: the mass of structure and residual strains in the car body after impact in the nine most dangerous points of the rear panel. The number of design variables in this problem is 13; that at face value suggests a large computational experiment (with several thousand tests). However, since it requires such a large amount of computational time to compute one vector of criteria, the optimization of design variables is very difficult to implement. Therefore, the initial model was decomposed in two approximate models of the bumper and the real panel. It was necessary to define the consistent solutions between subsystems of the bumper and the real panel. For this purpose, 300 tests were carried out. Each test required no more than 10 minutes. The analysis of the obtained solutions has shown that the prototype cannot be improved. As a result, the problem has been reformulated (i.e., the designs of the bumper and the rear panel have been modified by introducing additional stiffening ribs). Fifteen consistent design variable vectors were defined. For these vectors we have calculated all criteria using the original model. The number of feasible solutions satisfying all constraints of the structure was nine. The number of Pareto optimal solutions was five.
3. *Problem of the operation development of a truck* [2]. In terms of computational time, this problem falls in between the previous two prob-

lems. The computation of a vector of criteria based on a system of complex differential equations took approximately 3 minutes. In the identification phase of this problem, wc identified 16 parameters of the mathematical model using 65 adequacy criteria and defined to what extent the model corresponds to the real system. In the initial problem statement, 4,096 tests were conducted. Only seven solutions met the criteria constraints and thus entered the feasible solution set. After analyzing the obtained results, new boundaries of the design variables were defined. The same number of tests (4,096) was conducted. As a result, 11 more feasible solutions were obtained. On the basis of solution of the identification problem and the definition of the feasible solution set, the problem of optimization by 20 performance criteria was solved next. These criteria were divided into the following groups: (1) comfort, (2) durability, (3) load preservation, and (4) safety. Twenty parameters were varied. Optimization was aimed to improve the prototype. 4,096 tests were conducted and 21 solutions satisfied all constraints and entered the feasible solution set. The Pareto set consisted of 20 solutions.

In addition, we provide below three more relevant examples:

4. In the problems of improving the rear axle housing for a truck (5 performance criteria, 12 pseudo-criteria, and 7 design variables, Section 5.3) and improving the truck frame prototype (4 performance criteria, 3 pseudo-criteria, and 21 design variables, Section 5.4) 256 and 300 trials are performed correspondingly.
5. In the problem of vector identification of parameters of a spindle unit (Section 6.3), we considered dynamic and thermal mathematical models with 23 and 13 criteria adequacy correspondingly. 512 trials are carried out in each problem. In the problems of optimization of design variables of a spindle unit, 6 performance criteria were minimized and 256 trials were performed in each problem. (For the problems 4 and 5, calculation of one criterion vector requires up to 4 minutes.)
6. Chapter 9 discusses the application of the PSI method for the design of the L_1 flight control system implemented on the two turbine-powered dynamically scaled Generic Transport Model (GTM), which is part of the Airborne Subscale Transport Aircraft Research aircraft at the NASA Langley Research Center. The problem statement has been corrected twice. In the first iteration 14 criteria and pseudo-criteria

were investigated, 1,024 trails were performed, and it took approximately 1 minute to compute one vector of criteria. In the second iteration 16 criteria and pseudo-criteria were investigated, 512 trails were performed, and it took approximately 10 minutes to compute one vector of criteria.

In all cases, the results surpassing prototypes in all criteria or in the most important ones were obtained. It is important to note that in order to obtain these results the statements of the problems, as a rule, have been repeatedly corrected.

References

[1] Statnikov, R. B., et al., *MOVI 1.4 (Multicriteria Optimization and Vector Identification) Software Package,* Certificate of Registration, Registere of Copyright, U.S.A., Registration Number TXU 1-698-418, Date of Registration: May 22, 2010.

[2] Statnikov, R. B., and J. B. Matusov, *Multicriteria Analysis in Engineering,* Dordrecht/Boston/London: Kluwer Academic Publishers, 2002.

[3] Egorov, I. N., et al., "Multicriteria Optimization of Complex Engineering Systems from the Design to Control," *Journal of Machinery Manufacture and Reliability,* Russian Academy of Sciences, No. 2, 1998, pp. 10–20.

[4] Beknev, V. S., I. N. Egorov, and V. S. Talyzina, "Multicriteria Design Optimization of Multistage Axial Flow Compressor," *Proceedings of the 36th International Gas Turbine Congress,* Budapest, Hungary, 1991.

[5] Egorov, I. N., and G. V. Kretinin, "Search for Compromise Solutions of the Multistage Axial Flow Compessor's Stochastic Optimization Problem," *Proceedings of the 3rd ISAIF (International Symposium on Experimental and Computational Aerothermodynamics of Internal Flows III),* Beijing, China, 1996, pp. 112–120.

[6] Egorov, I. N., et al., "Problems of Design and Multicriteria Control for Adjustable Technical Systems," *Doklady Physics* (Russian Academy of Sciences), Vol. 43, No. 3, 1998, pp. 181–184, translated from *Doklady Akademii Nauk,* Vol. 359, No. 3, 1998, pp. 330–333.

[7] Statnikov, R. B., et al., "Multicriteria Analysis Tools in Real-Life Problems," *Journal of Computers & Mathematics with Applications,* Vol. 52, 2006, pp. 1–32.

[8] Wasserman, P. D., *Advanced Methods in Neural Computing,* New York: Van Nostrand Reinhold, 1993.

[9] Anderson, T. W., *An Introduction to Multivariate Statistical Analysis,* 3rd ed., New York: Wiley-Interscience, 2003.

[10] Chatterjee, S., and A. S. Hadi, "Influential Observations, High Leverage Points, and Outliers in Linear Regression," *Statistical Science,* Vol. 1, No. 3, 1986, pp. 379–416.

[11] Vapnik, V., *Statistical Learning Theory,* New York: John Wiley & Sons, 1998.

[12] Statnikov, R. B., et al., "Visualization Approaches for the Prototype Improvement Problem," *Journal of Multi-Criteria Decision Analysis,* No. 15, 2008, pp. 45–61.

[13] Bondarenko, M. I., et al., "Construction of Consistent Solutions in Multicriteria Problems of Optimization of Large Systems," *Physics-Doklady,* Vol. 39, No. 4, 1994, pp. 274–279, translated from *Doklady Rossiiskoi Akademii Nauk,* Vol. 335, No. 6, pp. 719–724.

8

Adopting the PSI Method for Database Search

8.1 Introduction

Database search engines are used in essentially all areas of modern life. There are two major types of information retrieval problems that are being solved by search engines. In the first type, a search engine matches a query of the client to a database and reports back results. For example, the client uses PubMed service (http://www.ncbi.nlm.nih.gov/pubmed/) to identify articles that talk about the BRCA1 gene in breast cancer using the query "BRCA1 breast cancer." Another example is when the client uses the Google search engine (http://www.google.com/) to identify a poem from the phrase "to whose immortal eyes." Usually, this type of problem can be solved efficiently and automatically (i.e., without interaction with the client) given that the database contains a sought solution.

In the second type of problem, the client uses a search engine to retrieve solutions (alternatives) from the database that are described by a set of contradictory characteristics. For example, such problems are widespread when searching for an airline ticket, a matching partner, or real estate on the Web. In a typical scenario, the database contains a very large number of alternatives described by a set of characteristics. Because of the large number of alternatives, the search engine allows the client to impose constraints on characteristics such that the resulting filtered set of alternatives satisfies all requests of the client. If these constraints are defined correctly, they determine so-called the feasible solution set where the most preferable solution should be then sought by the client. However, in the majority of cases, the client defines constraints intuitively and thus often incorrectly. As such, many interesting solutions that, under certain assumptions, could have been feasible will be lost. This results in the poor or even empty feasible solution set. This is especially the case when the client deals with many contradictory characteristics and many alternatives[1]. Current search engines do not provide tools for systematic construction of the feasible solution set. This is a major shortcoming of existing database search engines, because a search for the most preferable solution depends, to a large extent, on the definition of the constraints, and this insurmountable task is shifted to the client's shoulders.

The problem of database search involves many contradictory characteristics and a large number of alternatives and necessitates decision-making [1]. The ultimate results also critically depend on the quality of database [2], the type of constraints on characteristics [3], and other factors. An important feature of the database search problem is that filtering of the number of solutions via the construction of the feasible solution set should precede decision-making.

That is why main thesis of this chapter is that systematic construction of the feasible solution set has a fundamental value for the client. In this chapter, we modify the PSI method to allow the construction of the feasible solution set specifically for the problems of database search. We propose a modified PSI method called DBS-PSI (DBS is an acronym for "database search") that allows us to significantly improve the quality of the search results by explicitly constructing and analyzing the feasible solution set and identifying the most preferable solutions.

8.1.1 Characteristics of Alternatives: Criteria and Pseudo-Criteria

We define a criterion as the characteristic of the alternative[2] that is related to the alternative's quality by monotonous dependence. In other words, all other things being equal, the more (or the less) the value of the criterion, the better the alternative. Contrary to the criterion, a pseudo-criterion is not related to an alternative's quality by monotonous dependence [4]. Pseudo-criteria do not need to be optimized; only their constraints have to be satisfied. These constraints are determined either by known and generally accepted standards or by the client's preferences. In most cases, constraints on both criteria and pseudo-criteria are not rigidly set. An algorithm allowing us to determine these constraints is provided in [5]. Examples of criteria and pseudo-criteria are given next.

1. In rare cases, when there are a few alternatives and characteristics, this task may be realistic for the client.

2. Examples of alternatives in database search engines are airline tickets (travel search Web sites), women/men (online dating services), houses (real estate search Web sites), cars (online services to buy cars), and so on.

Let us consider four popular uses of database search engines and determine several criteria and pseudo-criteria to illustrate these definitions:

- Using travel search Web sites to buy airline tickets, for example, Kayak travel search engine (http://www.kayak.com). The criteria are cost, number of stops, and duration. The pseudo-criteria are takeoff time, landing time, airline, airport (see Figure 8.1). For example, the client most often specifies a range of constraints on takeoff time and landing time. However, if he or she wanted to depart/arrive as early/late as possible, takeoff time and landing time should be considered as criteria. The choice of destination airport may also be a criterion if the client has to hurry from the airport to work, and as a result he or she chooses the airport closest to his or her office.
- Using online dating services to look for a matching partner, for example, Yahoo! Personals (http://personals.yahoo.com/). The criteria are educa-

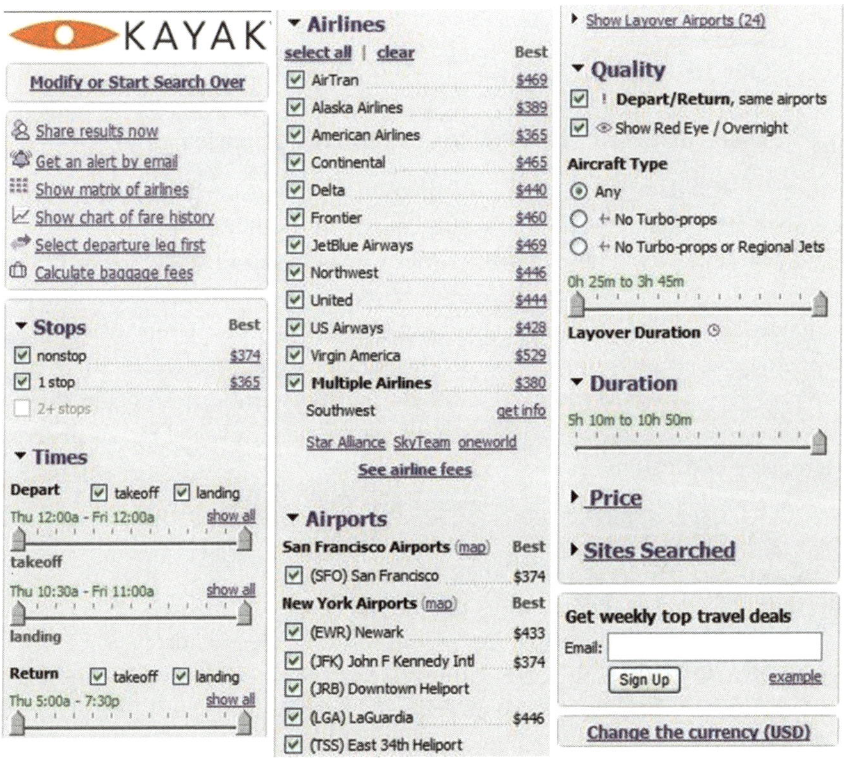

Figure 8.1 Kayak travel search engine. (Source: Available at http://www.kayak.com.) Fragments.

tion level and income level. The pseudo-criteria are appearance and lifestyle. Appearance contains the subsets of more specific pseudo-criteria such as ethnicity, height, body type, eye color, and hair color. Lifestyle contains the subset of more specific pseudo-criteria such as marital status, profession, religion, and interests. In general, when we talk about one's height, weight, and age, it generally makes no sense to say that the higher (lower) the values of these characteristics, the better the alternative. These characteristics must lie within certain boundaries.

- Using real estate search websites to buy a house, for example, Realtor.com (http://www.realtor.com/). The criteria are cost, distance to work, and school district (for school-age children). The pseudo-criteria are ceiling height and number of floors. If the client is handicapped and uses a wheelchair, the number of floors becomes a criterion.
- Using online services to buy cars, for example, Cars.com (http://www.cars.com/). The criteria are cost, fuel consumption, and operating cost. The pseudo-criteria are wear resistance, durability, and the operating life of car parts.

8.1.2 General Statement of the Problem and Solution Approach

Assume that a database contains N alternatives (solutions), each of which is described by k characteristics[3] (pseudo-criteria and criteria) $\mathbf{\Phi}_v^j$, for $v = 1,\ldots, k$ and $j = 1, \ldots, N$. These characteristics can be contradictory, so improving some characteristics leads to the deterioration of the others. Also assume that the database contains so many alternatives that the client experiences significant difficulties while analyzing them and choosing the most preferable alternative. The problem is how to filter the set of alternatives without losing any alternatives that are acceptable to the client. This can be accomplished by constructing the feasible and Pareto optimal sets.

We are given a vector of characteristics $\Phi = (\Phi_1, \Phi_2, \ldots, \Phi_k)$, where $\Phi_1, \Phi_2, \ldots, \Phi_c$ are criteria that we want to minimize and $\Phi_{c+1}, \Phi_{c+2}, \ldots, \Phi_k$ are pseudo-criteria. Alternatives $\mathbf{\Phi}^j$ that: (1) satisfy all criteria constraints, that is, $\mathbf{\Phi}_v^j$ (v = 1, ..., c) such that $\mathbf{\Phi}_v^j \le \mathbf{\Phi}_v^{**}$ (or $\mathbf{\Phi}_v^j \ge \mathbf{\Phi}_v^{**}$ in the case of maximization) and (2) satisfy constraints on pseudo-criteria, that is, $\mathbf{\Phi}_v^j$ (v = c + 1, ..., k) such that $\mathbf{\Phi}_v^{*} \le \mathbf{\Phi}_v^j \le \mathbf{\Phi}_v^{**}$, constitute the feasible solution set D. Here $\mathbf{\Phi}_v^{**}$ is the worst value of criteria which the client considers as acceptable; constraints

3. It is worth noting that often client can a priori state "rigid" requirements to certain characteristics that cannot be changed under any circumstances. In such cases, when searching for the most preferable solution, it is necessary to exclude alternatives that did not satisfy those requirements and then search for feasible solutions as described in this section.

on pseudo-criteria Φ_v^* and Φ_v^{**} correspond to the best and worst values, respectively.

We solve the general problem by first constructing the feasible solution set D by imposing constraints on these criteria and pseudo-criteria mentioned via dialogues of the client with the computer. Next, we define the Pareto optimal set $P \subseteq D$ using only criteria $\Phi_1, \Phi_2, \ldots, \Phi_c$. Recall that an alternative $\Phi^{j_0} \in D$ is called a Pareto optimal if there exists no alternative $\Phi^j \in D$ such that $\Phi_v^j \leq \Phi_v^{j_0}$ for all $v = 1, \ldots, c$ and $\Phi_{v_0}^j < \Phi_{v_0}^{j_0}$ for at least one $v_0 \in \{1, \ldots, c\}$. In other words, alternative Φ^{j_0} cannot be improved by all criteria $\Phi_1, \Phi_2, \ldots, \Phi_c$ simultaneously. A set $P \subseteq D$ is called a Pareto optimal set if it consists of only Pareto optimal alternatives. Finally, the client determines the alternative $\Phi^{j_0} \in P$, which is the most preferable among the alternatives belonging to the set P.

8.1.3 Motivation of the Problem Statement

Concessions Φ_v^* and Φ_v^{**} are something like a market where we have the opportunity to trade until we get what we need. Of course, not all concessions are possible, but that is outside the scope of our work. By investigating the set of Pareto solutions, the clients see what can and what cannot be achieved. They are able to choose the most preferable alternative in this set, $\Phi^{j_0} \in P$. At the same time, they know that there exist no better alternatives in the database. The Pareto optimal set plays an important role in vector optimization problems because it can be analyzed more easily than the feasible solution set and because the optimal alternative always belongs to the Pareto optimal set, irrespective of the system of preferences used by the client for comparing alternatives belonging to the feasible solution set.

8.1.3.1 Problem Considered by This Work

We primarily consider problems with the following features:

- Characteristics are contradictory.
- The dimensionality of the characteristics vector may reach dozens.
- The number of alternatives may reach many thousands.
- The constraints on these characteristics are often stringent.
- The constraints on characteristics Φ_v^* and Φ_v^{**} can be defined in an interactive mode, in the process of dialogues of the client with the computer.

8.1.3.2 What Do the Client and Search Engine Developers Need to Know?

Search engine developers must provide all necessary characteristics (criteria and pseudo-criteria). The client must understand which constraints on characteristics are acceptable and which are not, without differentiating between criteria and pseudo-criteria when constructing the feasible set. After determining the feasible set (feasible alternatives), the client must decide personally which one of these characteristics is a criterion and which one is a pseudo-criterion. This is because the same characteristics may be a criterion for one client and pseudo-criteria for another client. Next come the analysis of the obtained alternatives and the selection of the most preferable one.

8.2 DBS-PSI Method

8.2.1 DBS-PSI Method as a New Paradigm of a Database Search

We mentioned earlier difficulties in constructing the feasible solution set correctly. To do this, the client must see what concessions he or she should make on characteristics and what he or she will get in return[4]. We want to show the process of generating concessions to the client and to interactively engage him or her in it.

Recall that the PSI method was designed to help the client determine constraints on characteristics (criteria and pseudo-criteria) and thus to construct the feasible solution set. After determining the feasible solution set, the Pareto optimal set is constructed only with consideration of the criteria. Then the client will search for the most preferable solution on the Pareto optimal set. The new DBS-PSI method [5] is designed to be used for a database search, and it consists of three main stages, presented next.

Stage 1: Compilation of Analysis Tables Via Computer

In this stage, the client transforms the database into analysis tables[5] as follows. For each characteristic an analysis table is compiled so that the values of $\Phi_v^1,\ldots,\Phi_v^N$ are arranged in increasing order; that is,

$$\Phi_v^{i_1} \le \Phi_v^{i_2} \le \ldots \le \Phi_v^{i_N},\ v = 1,\ldots,k \tag{8.1}$$

4. We note that in search engines it is possible to revise constraints. However, the client is expected to do this at a very intuitive level.

5. The foundation of the PSI method is test tables. The principal difference between test tables and analysis tables is that in the former case, solutions are obtained with the use of generators (e.g., uniformly distributed sequences), and in the latter case solutions are obtained from a database. Also, the parameters are absent in the search engines that are discussed in this section.

where $i_1, i_2, \ldots, i_N$ are the numbers of alternatives (a separate set for each v). Taken together, the k tables form complete analysis tables. In analysis tables, the list of nonnumerical criteria is arranged in order of the client's preferences.

Stage 2: Selection of Characteristic Constraints

This stage includes the dialogue of a client with the computer. Let $\min\Phi_v$ and $\max\Phi_v$ be unfeasible solutions for the vth characteristic (pseudo-criterion and/or criterion), and let $\mathbf{\Phi}_v^*$ and/or $\mathbf{\Phi}_v^{**}$, respectively, be the minimum possible concessions from $\min\Phi_v$ and/or $\max\Phi_v$, ($\min\mathbf{\Phi}_v^* \leq \mathbf{\Phi}_v^j \leq \mathbf{\Phi}_v^{**} \leq \max\Phi_v$), where $[\min\Phi_v; \mathbf{\Phi}_v^*)$ and $(\mathbf{\Phi}_v^{**}; \max\Phi_v]$ are unfeasible solution domains. Then the remainder of table $[\mathbf{\Phi}_v^*; \mathbf{\Phi}_v^{**}]$ is the maximum search interval for feasible solutions. After examining the characteristics arranged in each of the analysis tables in increasing (decreasing) order, starting with $\min\Phi_v$ and/or $\max\Phi_v$, the client analyzes each value. The first feasible value found will be $\mathbf{\Phi}_v^*$ ($\mathbf{\Phi}_v^{**}$). If $\min\Phi_v$ and/or $\max\Phi_v$ are feasible values, there is no need to make the corresponding concessions. In general, we will denote the feasible solution search interval as $[\mathbf{\Phi}_v^*; \mathbf{\Phi}_v^{**}]$. If the selected values of $[\mathbf{\Phi}_v^*; \mathbf{\Phi}_v^{**}]$ are not a maximum, then many interesting solutions may be lost, since some of the characteristics are contradictory.

The client has to consider one characteristic (analysis table) at a time and specify the respective constraints $\mathbf{\Phi}_v^*$ and $\mathbf{\Phi}_v^{**}$. Then the client proceeds to the next analysis table, and so on. Note that the revision of the characteristic constraints does not cause any difficulties to the client.

It is important to emphasize that a human-computer dialogue is very convenient for the client: he or she does not have to change the values of some characteristics at the expense of others. He or she sees one analysis table and sets the appropriate constraints; then he or she repeats the same process with another table, and so on. After the client defines the constraints, the computer searches for the feasible solutions. For example, if the client has defined the constraints on the first and second characteristics, he or she immediately obtains information on feasible solutions based on these characteristics. After defining the constraints on the third characteristic, the client obtains feasible solutions based on three characteristics, and so on. In other words, the client sees what price he or she pays for making concessions on various characteristics (i.e., what the client loses and gains). Such an analysis gives the client valuable information on the advisability of revising various characteristic constraints with the aim of improving the basic characteristics. Thus, in this method, the client imposes constraints on each characteristic in succession and follows the construction of the feasible solution set step by step. Moreover, the client can go back at any time and reexamine the constraints depending on the feasible solution set obtained.

Stage 3: Verification of the Solvability of a Problem Via Computer

Let the client fix a characteristic (e.g., Φ_{v_1}), and consider the corresponding analysis table (8.1). Let S_1 be the number of values in the table satisfying the selected characteristic constraints:

$$\Phi^{*}_{v_1} \leq \Phi^{i_1}_{v_1} \leq \ldots \leq \Phi^{i_{S_1}}_{v_1} \leq \Phi^{**}_{v_1} \tag{8.2}$$

Then characteristic Φ_{v_2} is selected by analogy with Φ_{v_1} and the values of $\Phi^{i_1}_{v_2},\ldots,\Phi^{i_{S_2}}_{v_2}$ in the analysis table are considered. Let the table contain $S_2 \leq S_1$ values such that $\Phi^{*}_{v_2} \leq \Phi^{i_j}_{v_2} \leq \Phi^{**}_{v_2}$, where $1 \leq j \leq S_2$. Similar procedures are carried out for each characteristic. If at least one alternative can be found for which all characteristic constraints are valid simultaneously, then the feasible solution set is nonempty and our problem is solvable. Otherwise, the client should return to Stage 2 and make certain concessions on characteristic constraints Φ^{*}_{v} and Φ^{**}_{v}. The procedure is iterated until the feasible solution set is nonempty. Now the Pareto optimal set is constructed. This is done by removing those feasible alternatives that can be improved with respect to all characteristics simultaneously.

As a rule, the client makes more concessions on the less important characteristics. These concessions may allow him or her to obtain a substantial gain in the most significant characteristics.

Notice that the number of characteristics must be no less than necessary. The client always wishes to optimize not one, but all of the most important characteristics, many of which are antagonistic. The greater the number of characteristics taken into account, the richer the information obtained about alternatives. The DBS-PSI method allows us to consider as many characteristics as necessary.

8.3 Searching for a Matching Partner

The purpose of this example is to give readers an idea of how to construct feasible and Pareto optimal solutions on the basis of analysis tables. Earlier we mentioned the example of using online dating services to look for a matching partner. Let us show how a male client will search for a matching female partner using the DBS-PSI method. Suppose the search characteristics are Height (Φ_1), Weight (Φ_2), Age (Φ_3), Income (Φ_4) and Education (Φ_5). Let the client have access to a database with alternatives. According to Stage 1 of the DBS-PSI method, the client should construct analysis tables (see Figure 8.2). From Figure 8.2 it can be seen that the database contains a large number of alternatives[6] and the values of the characteristics vary over fairly large ranges: from a

6. For example, #12,089, #34,006, #95,175, #30,617, #88,945, and so on. Each of the women in the analysis tables is designated as an alternative #.

#	*Height, Φ_1*	#	*Weight, Φ_2*	#	*Age, Φ_3*	#	*Income, Φ_4*	#	*Education, Φ_5*
284	4'11"	5,298	99.2 lbs	971	25 years	1,021	$25,000	12,907	Some high school
…	…	…	…	…	…	…	…	…	…
689	5'2"	88	125.7 lbs	472	30	894	$25,000	9,705	High school graduate
…	…	…	…	…	…	…	…	…	…
27	**5'3"**	…	…	…	…	…	…	12,096	High school graduate
63	5'3"	**7,632**	**127.9 lbs**	…	…	…	…	63	Some college
…	…	97	130.1 lbs	…	…	14,007	$35,000	…	…
12,089	5'3"	…	…	888	34	7	$50,000	…	…
92	5'4"	894	130.1 lbs	**7,632**	**34**	…	…	…	…
…	…	…	…	…	…	**27**	**$50,000**	**7,632**	**Some college**
894	5'4"	10	132.3 lbs	9	35	80,092	$50,000	**27**	**College graduate**
…	…	…	…	…	…	21	$75,000	…	…
1,964	5'4"	1,093	138.9 lbs	23	36	…	…	…	…
38	5'5"	…	…	894	36	…	…	…	…
7,632	**5'5"**	…	…	…	…	…	…	…	…
30,617	5'5"	**27**	**143.3 lbs**	…	…	**7,632**	**$75,000**	…	…
…	…	…	…	**27**	**38**	245	$100,000	…	…
994	5'6"	…	…	…	…	…	…	…	…
…	…	30,617	150 lbs	300	40	6,005	$100,000	88,945	College graduate
…	…	…	…	…	…	711	$150,000	992	Postgraduate
…	…	…	…	…	…	…	…	…	…
76	6'11"	66	183 lbs	34,006	65	90,661	$150,000	95,175	Postgraduate

Figure 8.2 Fragment of analysis tables.

height of 4'11" (alternative #284) to 6'11" (alternative #76), from a weight of 99.2 lbs (alternative #5,298) to 183 lbs (alternative #66), from 25 years of Age (alternative #971) to 65 years of Age (alternative #34,006), from an Income of $25,000 (alternative #1,021) to $150,000 (alternative #90,661), and from an Education of Some High School (alternative #12,907) to Postgraduate (alternative #95,175). The more alternatives there are in the analysis tables, the greater the client's choice.

According to Stage 1, the client should construct search intervals for feasible solutions. After studying the analysis table for the first characteristic (Height), the client has determined the search interval for feasible solutions $[\Phi_1^*;\Phi_1^{**}]$ = [5'2"; 5'6"] (see Figure 8.2).

Alternatives #689, …, #27, …, #994 correspond to this interval. The client then proceeds to the second analysis table, to Weight, and determines the search interval for feasible solutions $[\Phi_2^*;\Phi_2^{**}]$ = [125.7 lbs; 150 lbs]. Alternatives #88, …, #7,632, …, #30,617 correspond to this interval. When switching from one analysis table to another (from Height to Weight), the computer records the total number of alternatives that satisfy the constraints on these characteristics simultaneously. As it turns out, there are four of these alternatives, #27, #894, #7,632 and #30,167. If there were no feasible alternatives, client should revise the constraints on the first and/or second characteristics. The client then proceeds to the third analysis table (Age) and constructs the interval $[\Phi_3^*;\Phi_3^{**}]$ = [30 years; 40 years] that is satisfied by alternatives #472, …, #888, …, #300. The total number of alternatives that satisfies the constraints on three characteristics simultaneously is reduced to three, #27, #894 and #7,632. Then the client examines Income (the fourth analysis table). Not a single alternative has satisfied the constraints \$150,000, and \$100,000. After setting the constraint to \$75,000, the alternative #7,632 is found that satisfies all constraints on four characteristics. Finally, the client reduces the constraint to \$50,000. As a result, the total number of alternatives that satisfy constraints on four characteristics is two, namely, #27 and #7,632 (see Figure 8.2).

In the last analysis table (the "Education" characteristic), not a single alternative satisfied the "Postgraduate" constraint. After setting the constraint to "Some College," two feasible alternatives (matching partners) #27 and #7,632 are obtained (shown with bold in the Figure 8.2). These alternatives satisfy all constraints on five characteristics. The values of characteristics of matching partners #27 and #7,632 also are given in Table 8.1. All constraints and matching partners were determined in an interactive mode.

In Figure 8.2, the areas of search for feasible alternatives are highlighted in tan (pseudo-criteria) and in light green (criteria) accordingly. Since our example is concerned with the construction of feasible solutions #27 and #7,632 only, some alternatives and their corresponding values of criteria have been replaced with "…" in the analysis table.

8.3.1 Conclusions of the Example

Constructing and studying analysis tables allowed the client to obtain two feasible alternatives. After that the client may wish to determine the optimal alternative set with consideration of all or the most important criteria. For example, if the client had searched for the optimal solutions according to the criterion

Table 8.1
Input Data and Some Search Results

Characteristics	Height, Φ_1	Weight, lbs, Φ_2	Age, years, Φ_3	Income, $, Φ_4	Education, Φ_5
Range of characteristic values	4′11″–6′11″	99.2–183	25-65	25,000–150,000	Some high school, some college, college graduate, postgraduate
Intervals of search for feasible solutions	5′2″–5′6″	125.7–150	30–40	50,000–150,000	Some college, college graduate, postgraduate
Feasible Alternatives (matching partners)					
#27	5′3″	143.3	38	50,000	College graduate
#7,632	5′5″	127.9	34	75,000	Some college

Income, the search engine would have given him alternative #7,632; alternative #27 would be better for the criterion Education. If two criteria, Income and Education, are considered simultaneously, the Pareto optimal solution set will contain the two specified solutions.

However, according to Yahoo! Personals (http://personals.yahoo.com/), upon starting a search, the client should impose constraints on age and geographical area (see Figure 8.3). In our opinion, these recommendations are incorrect, since the client can lose compromise solutions that are acceptable to him. These constraints should be defined in the process of studying the analysis tables; after finding the feasible alternatives (matching partners) in an interactive mode, the remaining solutions will automatically become unfeasible.

8.4 Summary

It is often the case that database search involves many contradictory criteria and many alternatives, and it is necessary to search for compromise solutions. This

Figure 8.3 Yahoo! Personals search engine. (Notes: available at http://personals.yahoo.com/.) Fragment.

can be efficiently accomplished by construction and analysis of the feasible solution set. However, current search engines do not systematically construct the feasible solution set, thus "hiding" many potentially interesting solutions from the client. To address this problem, we introduced the DBS-PSI method, which allows the client to:

1. Construct analysis tables.
2. Determine the feasible solution set in the process of dialogues with the computer.
3. Find the Pareto optimal set.
4. Identify the most preferable solution.

The process of searching for the feasible solution set based on analysis tables is client-friendly: clients see what they lose and what they gain for every concession. The DBS-PSI method as a new paradigm of database search can be used in all areas of modern life and promises to increase efficiency of modern search engines. The relevancy of the proposed methodology is apparent given ongoing expansion of the World Wide Web and search engines.

References

[1] Saaty, T., "Decision Making with the Analytic Hierarchy Process," *International Journal of Services Sciences,* Vol. 1, No. 1, 2008, pp. 83–98.

[2] Li, X., et al., "Problems and Systematic Solutions in Data Quality," *International Journal of Services Sciences,* Vol. 2, No. 1, 2009, pp. 53–69.

[3] Stefansen, C., and S. E. Borch, "Using Soft Constraints to Guide Users in Flexible Business Process Management Systems," *International Journal of Business Process Integration and Management,* Vol. 3, No. 1 2008, pp. 26–35.

[4] Sobol', I. M., and R. B. Statnikov, *Selecting Optimal Parameters in Multicriteria Problems,* 2nd ed., (in Russian), Moscow: Drofa, 2006.

[5] Statnikov, R. B., et al., "DBS-PSI: A New Paradigm of Database Search," *International Journal of Services Sciences,* Vol. 4, No. 1, 2011, pp. 1–13.

9

Multicriteria Analysis of L_1 Adaptive Flight Control System

9.1 Objective of the Research

Inner-loop adaptive flight control systems[1] are seen as an appealing technology [1] to improve aircraft performance with reduced pilot compensation in adverse flight conditions or in the event of control surface failures and vehicle damage. Under these conditions, which are characterized by a high degree of uncertainty with respect to a nominal aircraft model, the achievable levels of performance and flying qualities (FQ) that a nonadaptive flight control system (FCS) can provide might be limited. However, in applications with stringent performance and robustness specifications [2], several limitations of conventional adaptive control architectures have been identified that render the design of these conventional adaptive controllers overly challenging. Among these limitations [2, 3], are of special relevance: (1) the lack of transient characterization of the closed-loop response, (2) the limited analysis framework for robustness and performance guarantees for closed-loop adaptive systems, and (3) the lack of systematic design guidelines to solve the trade-off between adaptation, performance, and robustness.

The L_1 adaptive control theory [4] appeared recently as a method for the design of robust adaptive control architectures using fast estimation schemes, with the potential to overcome the limitations mentioned above. The key feature of L_1 adaptive control is the decoupling of adaptation and robustness. In fact, in L_1 adaptive control architectures, the speed of adaptation is limited

1. Chapter 9 is written in collaboration with Enrick Xargay (University of Illinois at Urbana-Champaign) and Vladimir Dobrokhodov (Naval Postgraduate School).

only by the available hardware (computational power and high-frequency sensor noise), while the trade-off between performance and robustness can be addressed via conventional methods from classical and robust control. Fast adaptation, which allows for the compensation of the undesirable effects of rapidly varying uncertainties and significant changes in system dynamics, is critical towards achieving guaranteed transient performance without enforcing persistency of excitation, applying gain scheduling of the control parameters, or resorting to control reconfiguration or high-gain feedback. Moreover, the systematic design procedures of the L_1 adaptive control theory significantly reduce the tuning effort required to achieve a desired closed-loop performance, which translates into a reduction in both the design cycle time and the development costs. With these features, the L_1 adaptive control architectures provide a suitable framework for the development of advanced flight critical systems.

In this chapter, we present the preliminary results of the application of the PSI method for the design optimization of the L_1 flight control system implemented on the GTM AirSTAR [5, 6] aircraft's current primary flight test vehicle, the GTM tail number T2. The T2 is a twin-engine jet-powered and dynamically-scaled (5.5%) civil transport aircraft, designed and instrumented to perform control law evaluation, experiment design and modeling research, in-flight failure emulation, and flight in upset conditions. The research control law developed for the NASA AirSTAR flight test vehicle has a primary objective of achieving tracking for a variety of tasks with guaranteed stability and robustness in the presence of uncertain dynamics, such as changes due to rapidly varying flight conditions during standard maneuvers, and unexpected failures. Ideally, the flight control system should provide level 1 flying qualities under nominal as well as adverse flight conditions.

The L_1 flight control system (L_1 FCS) used for this application is a three-axis angle of attack (α), roll rate (p), and sideslip angle (β) all-adaptive flight control system. The FCS consists of two decoupled L_1 controllers, one for the longitudinal channel and another one for the control of the lateral-directional dynamics. On one hand, the longitudinal L_1 controller is implemented as a single-input single-output controller and uses feedback in angle of attack and pitch rate to generate an elevator control signal in order to track angle of attack reference signals; on the other hand, the lateral/directional L_1 controller is a multiple-input multiple-output architecture and uses feedback in sideslip angle, roll rate, and yaw rate to generate aileron and rudder commands in order to track sideslip-angle and roll-rate reference signals with reduced coupling. In the current L_1 FCS, the pilot directly adjusts the thrust level using the throttle lever. The reader is referred to [7–9] for a more detailed explanation of the L_1 FCS implemented on the NASA AirSTAR flight test vehicle.

The main challenge for the design of the L_1 FCS is the optimal tuning of its elements to provide desired flying qualities with satisfactory robustness margins. While the theory of L_1 adaptive control provides partial systematic design guidelines to address the trade-off between performance and robustness, the optimization of the design of the L_1 adaptive controller is still largely open and hard to address. The main difficulty is the nonconvex and nonsmooth nature of the underlying optimization problem that involves the L_1 norm of cascaded linear systems. Randomized algorithms have been proven to be useful in control-related nonconvex optimization problems, and therefore they appear as appealing methods for the optimal design of L_1 adaptive controllers [4, 10].

The optimization design methodology proposed in this chapter unfolds in three sequential steps. First, we take advantage of the systematic design guidelines of the L_1 adaptive control theory to find a nominal prototype solution satisfying a given set of control specifications. Then, taking the prototype solution as a reference design, the PSI method is used for the construction of the feasible solution set and for determining an initial direction of improvement for the design of the flight control system [11–13]. Finally, the PSI method is again used to determine an optimal design that satisfies performance and robustness constraints and improves the prototype design. In particular, this design methodology demonstrates the suitability of the PSI method as a tool for formulating and solving multicriteria optimization problems for the design of adaptive FCS and also illustrates that the consistent application of the systematic design guidelines of L_1 adaptive control becomes particularly beneficial for the construction of the feasible solution set.

Following this approach, we next present preliminary results on the application of the PSI method and the MOVI software [12] to the design optimization of the L_1 FCS for the NASA GTM. We notice that, although both the PSI and the software package MOVI were specially developed to address problems with design variable and the criteria spaces of high dimensionality, for the sake of clarity in the presentation and discussion of the results, we keep the design problem within a reasonable complexity, and the design procedure is applied to the design of the longitudinal channel only. Finally, we also notice that the results included in this study are obtained by the MOVI software package combined with the MATLAB environment and are based on the full nonlinear simulation of the two-engine-powered, dynamically-scaled GTM AirSTAR system tail number T2, which was released by NASA in December 2009 [14].

The formulation of the design optimization of the L_1 FCS of the NASA AirSTAR flight test vehicle is addressed in Section 9.2. In particular, this section provides a detailed discussion of the workflow of the optimization process, including the identification of a nominal prototype design, the construction of the feasible solution set, and the improvement of the prototype. Finally, Section 9.3 provides solution of the optimization problem and summarizes the key results and presents the main conclusions.

9.2 Prototype: Criteria and Design Variables

The design of the longitudinal L_1 FCS is based on the linearized short-period dynamics of the GTM at the reference flight condition (80 kt, 1,000 ft). The L_1 FCS with its main elements and the design variables (*DV*) is represented in Figure 9.1 with a following corresponding explanation of the control architecture. The motivated choice of the *DVs* will be given in the following section. Since the airplane is level 1 FQ at this flight condition[2], the desired dynamics of the state predictor are chosen to be close to those of the actual aircraft.

For the nominal prototype design, the natural frequency of the poles of the system is reduced from $6\frac{\text{rad}}{\text{sec}}$ to $5.5\frac{\text{rad}}{\text{sec}}$, while the damping ratio is increased from 0.47 to 0.85. A first-order lowpass filter with DC gain 1 and a bandwidth of $20\frac{\text{rad}}{\text{sec}}$ was used in the matched contribution to the elevator command, while two cascaded first-order lowpass filters were used in the unmatched channel, both having DC gain equal to 1 and bandwidths of $5\frac{\text{rad}}{\text{sec}}$ and $7\frac{\text{rad}}{\text{sec}}$, respectively. Finally, the adaptation sampling time was set to $\frac{1}{600}$ sec, which corresponds to the minimum integration step allowed in the AirSTAR flight control computer. A first-order prefilter with $20\frac{\text{rad}}{\text{sec}}$ of bandwidth was added to shape the pilot command. This prototype design of the state predictor,

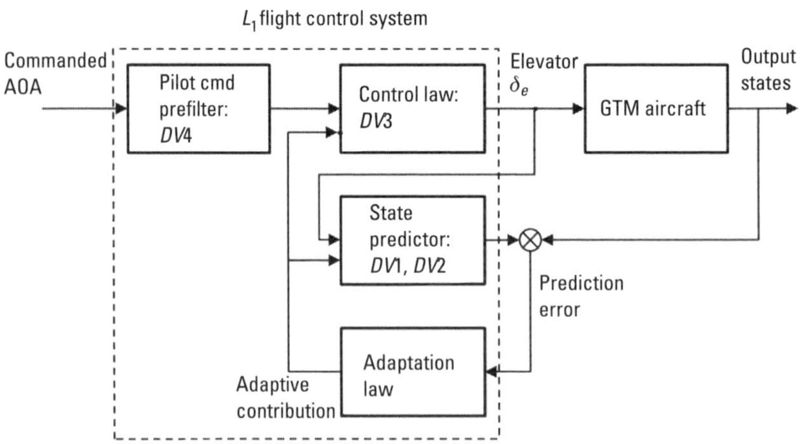

Figure 9.1 Longitudinal channel of the L_1 flight control architecture.

2. This FQ rating is based on offset-landing tasks performed in the AirSTAR real-time simulator at NASA LARC in the absence of atmospheric turbulence.

the lowpass filters, the adaptation sampling rate, and the prefilter delivers the angle of attack α (AOA) response similar to the desired one; see Figure 9.2 where α_{des} designates the desired angle-of-attack response.

It is worth noting here that this nominal design ensures a time-delay margin of the inner loop of approximately 85 ms and a gain margin of 7.2 dB, in wings-level flight at the flight condition of 80 knots and 1,000 ft. At this flight condition, the FQs are predicted to be level 1 and the FCS design has no predicted pilot induced oscillation (PIO) tendencies (for an acquisition time of 1.5 seconds). Naturally, these metrics are the performance criteria, and their initial values provided above for the nominal design will serve as the guidelines for the motivated definition of the criteria constraints.

We next present preliminary results of the application of the PSI method and the MOVI software to the design optimization of the L_1 FCS for the NASA GTM airplane [11–13]. In particular, the objective of the optimization task is to minimize the difference between the desired and actual AOA responses, while ensuring satisfactory FQs and not overloading the elevator actuators. For this purpose, a number of criteria reflecting the differences were formulated. The corresponding discussion is presented in the following sections.

In what follows, we provide a description of the design variables alone with the performance and robustness metrics used for this study and obtain a prototype design that achieves desired flying qualities with satisfactory robustness margins. Then we utilize the capability of the PSI method and the software package MOVI to improve the prototype design.

9.2.1 Design Variables

Next we define the set of design parameters of the flight control architecture considered for optimization. Since the primary objective is to improve the flying qualities of the prototype design while guaranteeing satisfactory robustness margins, we include the natural frequency and the damping ratio of the poles of the state-predictor dynamics (which can speed up or slow down the response of the augmented aircraft), and the bandwidth of the lowpass filter in the matched channel (which can be used to adjust the time-delay margin of the inner loop) as optimization parameters; see Figure 9.1 illustrating their place in the L_1 flight control architecture. In addition, we also consider the optimization of the bandwidth of the prefilter, which can be used to shape the pilot command as to prevent elevator rate limiting and avoid structural mode interaction. The following list summarizes the set of optimization parameters that define the design variable space:

- *DV1:* Natural frequency of the state-predictor poles (rad/sec);
- *DV2:* Damping ratio of the state-predictor poles;

- *DV3:* Bandwidth of the "matched" lowpass filter (rad/sec);
- *DV4:* Bandwidth of the pilot-command prefilter (rad/sec).

Therefore, the design variable vector is $DV = \{DV1, DV2, DV3, DV4\}$. We notice that, consistent with the theory of L_1 adaptive control, the adaptation sampling time is set to match the minimum integration step allowed in the AirSTAR flight control computer, and therefore it is not included as an optimization parameter.

9.2.2 List of Criteria and Pseudo-Criteria

The set of design criteria considered in this study is chosen to evaluate performance and robustness properties of the GTM aircraft augmented with the L_1 FCS. To provide an adequate assessment of the performance characteristics and flying qualities of the L_1-augmented aircraft both pilot-off-the-loop and pilot-in-the-loop performance metrics are included in the design procedure. The metrics considered can thus be classified in three categories:

1. Pilot-off-the-loop performance metrics;
2. Robustness metrics;
3. Flying qualities and PIO metrics.

Because the present material addresses only the design of the longitudinal channel of the L_1 FCS, the set of metrics used in this study are mainly based on the (time-domain) longitudinal response of the GTM with the L_1 FCS closing the inner loop. Further, we provide a detailed description of the specific metrics used for the design improvement of the adaptive control system. We note that some of the metrics used in this study were also proposed in [15] for the evaluation of aircraft augmented with an adaptive FCS.

9.2.2.1 Pilot-off-the-Loop Performance Metrics

This first set of metrics evaluates the pilot-off-the-loop performance of the augmented aircraft by characterizing its response to step inputs. In particular, the pilot-off-the-loop performance metrics are based on the time-domain response of the augmented aircraft to a step command of 3° held for 4 seconds in the AOA (see Figure 9.2), starting from a wings-level flight condition. The metrics capture the deviation of the actual response of the aircraft from a given desired response, which is defined to provide satisfactory flying qualities without reaching the physical limits of the platform, as well as different measures of control activity, load factor, and cross-coupling.

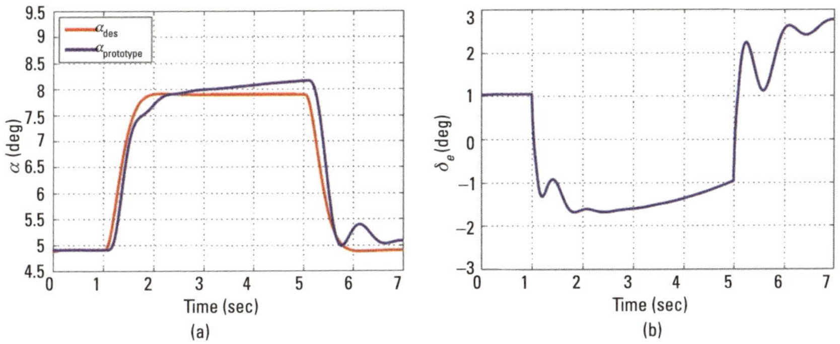

Figure 9.2 Prototype design, 3°-AOA step response (a) Angle of attack, α; (b) elevator deflection, δ_e.

Next we provide a description of the metrics included in the study. First, however, we need to introduce some key notation to facilitate the definition of these metrics. Below, alone with the previously defined α (AOA) and α_{des}, the α_{cmd} is the angle-of-attack pilot command; β is the angle of sideslip; β_{des} is the sideslip-angle desired response; p is the roll rate; p_{des} is the roll-rate desired response; A_z is the vertical acceleration; δ_e is the elevator deflection command; and $\tilde{x}(t)$ is the L_1 state predictor error.

Let t_0 be the time instant at which the step command is applied, and define t_f as the final time instant considered for the performance evaluation ($t_f = t_0 + 4$ seconds). With these notations, the metrics are defined as follows:

- *P1, final deviation:* This metric captures the final deviation of the actual AOA response from the desired AOA response at 4 seconds after the application of the step command. This metric is set to zero if the actual response reaches the AOA reference command before the end of the 4-second step:

$$P1 = \begin{cases} \left|\alpha(t_f) - \alpha_{des}(t_f)\right| & \text{if } \alpha(t) < \alpha_{cmd}, \forall t \in [t_0, t_f] \\ 0 & \text{otherwise} \end{cases}$$

  This metric can be used to penalize or exclude sluggish responses. In this study this metric is normalized to the amplitude of the step command (3°).

- *P2, maximum deviation from desired AOA response:* This metric captures the maximum deviation (in absolute value) of the actual AOA response from the desired AOA response:

$$P2 = \max_{t \in [t_0, t_f]} \left| \alpha(t) - \alpha_{des}(t) \right|$$

This metric can be redefined in terms of normalized AOA responses.

- *P3, integral deviation from desired AOA response:* This metric is defined as the (truncated) L_2 norm of the deviation of the actual AOA response from the desired AOA response:

$$P3 = \int_{t_0}^{t_f} \left| \alpha(t) - \alpha_{des}(t) \right| dt$$

Similar to *P*2, this metric can be redefined in terms of normalized AOA responses.

- *P4, overshoot in AOA response:* This metric captures possible overshoots and low-damping characteristics in the AOA response:

$$P4 = \begin{cases} \max_{t \in [t_0, t_f]} \left| \alpha(t) \right| \text{if}\, \alpha(t) > \alpha_{cmd}, \text{ for some } t \in [t_0, t_f] \\ \alpha_{cmd} \qquad \text{otherwise} \end{cases}$$

Similar to *P*1, this metric can be normalized to the amplitude of the step command (3°).

- *P5, maximum deviation from desired AOA rate response:* This metric captures the maximum rate deviation (in absolute value) of the actual AOA response from the desired AOA response:

$$P5 = \max_{t \in [t_0, t_f]} \left| \dot{\alpha}(t) - \dot{\alpha}_{des}(t) \right|$$

Similar to *P*2 and *P*3, this metric can be redefined in terms of normalized AOA responses.

- *P6, integral deviation from desired AOA rate response:* This metric is defined as the (truncated) L_2 norm of the rate deviation of the actual AOA response from the desired AOA response:

$$P6 = \int_{t_0}^{t_f} \left| \dot{\alpha}(t) - \dot{\alpha}_{des}(t) \right| dt$$

Similar to *P2*, *P3*, and *P5*, this metric can be redefined in terms of normalized AOA responses.

The metrics *P1* to *P6* provide a good first characterization of the transient response of the augmented aircraft when compared to a given desired response. Next, we present a set of metrics that can be extracted from the same step response experiment utilizing different flight dynamics characteristics which complement the AOA-based metrics defined above.

- *P7, maximum vertical acceleration:* Load factor (and passenger comfort) requirements can be captured by the maximum vertical acceleration during the step response:

$$P7 = \max_{t \in [t_0, t_f]} |A_z(t)|$$

This metric can also be normalized to the amplitude of the step command (3°).

- *P8, control effort:* This metric is defined as the (truncated) L_2 norm of the elevator deflection command:

$$P8 = \int_{t_0}^{t_f} |\delta_e(t)| \, dt$$

Similar to *P1*, *P4*, and *P7*, this metric can be normalized to the amplitude of the step command (3°). This metric can be used to penalize flight control designs that require a high control activity to achieve a desired control objective. It is important to note, however, that a high control effort might just be the result of a faster AOA response, and therefore a large P8 might not always be an undesirable response characteristic.

- *P9, maximum elevator rate:* Excessive control rate can be identified by the following metric:

$$P9 = \max_{t \in [t_0, t_f]} |\dot{\delta}_e(t)|$$

This metric, which can also be normalized to the amplitude of the step command (3°), can be used to penalize designs with high elevator rates in order to prevent undesirable effects from rate limiting.

- *P10, maximum elevator acceleration:* High-order derivatives of the control commands are coupled to the flexible modes of the aircraft. The following metric, based on the second derivative of the elevator command, can be used to capture excessive accelerations and oscillations in the control command that could potentially lead to unwanted structural mode interactions:

$$P10 = \max_{t \in [t_0, t_f]} \left| \ddot{\delta}_e(t) \right|$$

This metric can also be normalized to the amplitude of the step command (3°).

- *P11, maximum of L_1 prediction error:* This metric captures the maximum error between the actual system state and the state of the L_1 state predictor, usually denoted by $\tilde{x}(t)$:

$$P11 = \max_{t \in [t_0, t_f]} \left\| \tilde{x}(t) \right\|_\infty$$

In L_1 adaptive control architectures, the accurate estimation of system uncertainties and the performance guarantees rely on the (small) "size" of the prediction error $\tilde{x}(t)$. This metric can thus be used to monitor the correct functioning of the L_1 adaptive controller.

- *P12, maximum deviation in cross-coupling dynamics:* This metric captures the lateral-directional coupling induced by a command in the longitudinal channel:

$$P12 = \max_{t \in [t_0, t_f]} \left| \delta_e(t) \right| \left(\left(\beta(t) - \beta_{des}(t) \right)^2 + \left(p(t) - p_{des}(t) \right)^2 \right)$$

We notice that this metric, which can also be normalized to the amplitude of the step command (3°), provides valuable information for the design of the lateral-directional FCS, rather than the longitudinal FCS. In fact, for the design of the longitudinal L_1 FCS, this metric provides little information and it would be more convenient to analyze the coupling in the AOA response induced by a command either in roll rate or sideslip angle. The analysis of the response of the system to commands in the lateral-directional channel and the design of the lateral-directional L_1 FCS are left (deliberately) for future work.

- *P13, integral deviation in cross-coupling dynamics:* This metric is the integral version of the previous cross-coupling metric and is defined as follows:

$$P13 = \int_{t_0}^{t_f} \left|\delta_e(t)\right| \left(\left(\beta(t) - \beta_{des}(t)\right)^2 + \left(p(t) - p_{des}(t)\right)^2 \right) dt$$

Similar to $P12$, this metric would be more adequate for the design of the lateral-directional control system, and it is included in this study only to illustrate a set of additional metrics that can be derived from the response of the augmented aircraft to a command in the longitudinal channel.

9.2.2.2 Robustness Margins

In this preliminary study, the only robustness metric considered for optimization is the *time-delay margin* of the closed-loop adaptive system defined at the input of the aircraft (time delay inserted at the elevator deflection command), and it is also derived from the time-domain response of the augmented aircraft. For a given wings-level flight condition and with the pilot-off-the-loop, a small perturbation in the trim (initial) condition is introduced. The time-delay margin is determined as the minimum time delay that produces sustained oscillations in the AOA response as the L_1 FCS tries to stabilize the aircraft at the given trim condition. In this chapter, this robustness metric will be denoted by $R1$.

We notice that the time delay introduced in the elevator control channel is in addition to any time delay that is already modeled in the AirSTAR simulation environment, which amounts approximately to 25 ms.

9.2.3 Criteria Addressing FQ and PIO Characteristics

Finally, predictions for both flying qualities and PIO tendencies have also been included in order to complement the pilot-off-the-loop performance metrics presented above. For this study, we consider the time-domain Neal-Smith (TDNS) flying qualities and PIO criteria, which were specifically developed for nonlinear aircraft dynamics and nonlinear FCS. The TDNS criterion is the counterpart in the time domain of the frequency-domain Neal-Smith criterion, and it is based on a step-tracking task with different acquisition-time requirements. For a detailed description of this criterion, the reader is referred to [16]. The reader can also find in [17] a study on the prediction of flying qualities and adverse pilot interactions in the GTM augmented with the L_1 FCS.

We use four different metrics, extracting all of them from the TDNS criterion for an acquisition time of 1.5 seconds, to characterize the FQ and PIO tendencies of the augmented aircraft:

- *FQ1, tracking performance:* In the TDNS criterion, the root-mean-squared tracking error is used to evaluate the closed-loop performance with the pilot in the loop. A value of zero means that the pilot is able to perfectly track (with zero error) the reference command after the specified acquisition time.
- *FQ2, pilot workload:* In the TDNS criterion, the pilot workload is given by the pilot compensation phase angle (in degrees), which is derived from the optimal pilot model obtained from the criterion. A value of zero means that there is no need for either pilot lead compensation or lag compensation.
- *FQ3, FQ level:* The two metrics *FQ*1 and *FQ*2 are used to determine the predicted *FQ* level based on the *FQ* boundaries proposed in the criterion. *FQ*3 is a discrete metric, and it only admits the values 1, 2, and 3, which correspond to level 1, level 2, and level 3 flying qualities, respectively.
- *FQ4, PIO tendency:* The TDNS criterion also provides a prediction for the susceptibility of the augmented aircraft to PIO. This PIO-susceptibility metric is used to complement the flying qualities metrics discussed above. According to the TDNS criterion, a value above 100 implies that the augmented aircraft is PIO-prone, whereas a value below 100 indicates a PIO-immune configuration.

The set of FQ metrics is used at the second step when improving a prototype design of the longitudinal channel of the L_1 FCS. For the first step of the prototype design and the exploration of the feasible set, only a subset of these metrics is utilized. The full set of metrics is used in the last stage to optimize the design of the adaptive control system.

Based on the objectives of the task and previous flight control design expertise the following vectors of criteria {*P*1, *P*2, *P*3, *P*4, *P*5, *P*6, *FQ*1, *FQ*2, *FQ*3, *FQ*4, *R*1} and pseudo-criteria {*P*7, *P*8, *P*9, *P*10, *P*11, *P*12, *P*13} were defined.

9.2.4 Criteria Constraints

Based on the metrics defined, the final design of the L_1 FCS should ideally verify the set of control objectives at the reference flight condition of 80 knots of (equivalent) airspeed and 1,000 ft of altitude. Corresponding to this flight conditions a set of three criteria constraints were defined a priori:

$$P1 \leq 0.1, FQ3 = 1 \text{ and } P4 \leq 1.2$$

The first and second conditions address directly the control specifications, namely, the final value of the step response within 10% of the desired, and the predicted level 1 FQ. The third inequality imposes a 20% constraint on the overshoot in the step response, establishing thus a (loose) bound on the acceptable transient performance characteristics of the actual AOA response. Due to the significant difficulty of defining all criteria constraints consistent with the feasibility of the solution, the rest of the constraints were identified interactively while analyzing the test tables.

9.3 Solutions and Analysis

At this point the design optimization task and the nominal prototype solution were completely defined. This section presents last two steps of iterative application of the PSI method to the design improvement of the longitudinal channel of the L_1 FCS. It is critical to mention that the objective of the first iteration is achieved utilizing a limited number of criteria consistent with the control metrics used by the L_1 synthesis procedure. Numerical implementation of this first step is relatively efficient with the "computational price" of one solution measured in minutes. At the second step, while utilizing an extended set of criteria involving complex numerical procedures, the efficiency of numerical implementation becomes critical; the "computational price" of one solution at this step is measured in tenth of minutes.

9.3.1 First Iteration

For the purpose of constructing the feasible solution set, we consider the following initial intervals of design variables, which have been defined to include the initial prototype vector (see Table 9.1).

This first step is used to find an initial direction of improvement for the design of the adaptive FCS. More precisely, we aim here at determining tight intervals for the design variables characterizing the state predictor (*DV*1 and *DV*2) that would provide level 1 *FQ* and would not deviate significantly from the desired response defined previously. To this end, the design is to be minimized with respect to the following reduced number of criteria {*P*1, *P*2, *P*3, *P*4, *P*5, *P*6, *FQ*1, *FQ*2}, which is a subset of the metrics described earlier.

The robustness metric *R*1 and the PIO metric *FQ*4 are not included in this first step because their evaluation is computationally expensive; these

Table 9.1
Initial Intervals of Design Variables

Design Variable	Prototype	Initial Intervals of Variation of Design Variables Min	Max
DV1	5.50E+00	4.00E+00	8.00E+00
DV2	8.50E-01	5.00E-01	1.10E+00
DV3	2.00E+01	5.00E+00	3.00E+01
DV4	2.00E+01	1.00E+01	5.00E+01

metrics will be considered in the next step of the optimization process when the domain of the design variables becomes significantly refined.

The metrics *P7* through *P10* are not included in the set of criteria to be minimized because improved flying qualities may require "high" values of these metrics. Nevertheless, including these metrics in the optimization process as pseudo-criteria may provide useful information regarding the dynamics of the augmented aircraft. They also provide additional performance characteristics of the set of Pareto optimal solutions. In particular, we note that the metric *P9*, which is subject to "soft" control specifications, is included as a pseudo-criterion (see [11, 13] for a detailed justification of this choice). Similarly, the metric *P11*, which can be used to monitor the correct operation of the L_1 adaptive controller, does not need to be minimized as long as it remains a couple of orders of magnitude below the system state (truncated) L_∞ norm. Finally, as explained previously, the metrics *P12* and *P13* are included for the sake of completeness and should be considered only for the design of the lateral-directional control system. Therefore, they are not included as criteria in the current task.

Next we present the results obtained in this first iteration of the optimization process, which are based on 1,024 tests [11–13]. Out of these 1,024 tests, 427 vectors did not satisfy the a priori given criteria constraints. The solutions that did not satisfy the constraints entered the table of criteria failures [12]. In particular, 160 design variable vectors failed the 20% overshoot criterion constraint on *P4*, and 267 constraints failed the level 1 *FQ* criterion constraint *FQ3*, while the 10% overshoot constraint on *P1* was not effective, therefore not rejecting any solutions. Analysis of the design variable vectors contained in the table of criteria failures indicates that: (1) solutions characterized by low natural frequencies and high damping ratios of the state-predictor poles ($DV1 < 5$ and $0.95 < DV2$) fail the level 1 *FQ* criterion constraint *FQ3*; and (2) solutions with very high natural frequencies of the state-predictor poles ($7.6 < DV1$) or solutions with high natural frequencies and low damping ratios of the state-predictor poles ($7 < DV1 < 7.6$ and $DV2 < 0.6$) fail the overshoot criterion constraint *P4*. The first set of design variable vectors leads to slow responses of the augmented aircraft to pilot commands, which require moderate or considerable pilot compensation and thus result in a predicted level 2 *FQ* that is inadmissible. The second set corresponds to the solutions providing fast responses to step commands with big overshoots, which differ significantly from the desired response for the augmented aircraft.

The remaining 597 vectors that did satisfy a priori given criteria constraints were used to construct the test table. While tightening the criteria constraints in the test table, the following criteria constraints were formulated (see Table 9.2). Note that while analyzing the test table, the constraint of *P*4 was significantly tightened to the value of 1.02. It is also worth mentioning that the response on criteria *P*1 is not presented in the table because all solutions of the test table provided identical response, *P*1 = 0.

Only 20 solutions were found to be feasible according to these criteria constraints, all of them contributing to the Pareto optimal solutions. A fragment of the criteria table is given in Table 9.3.

Analysis of the criteria table shows that solution #993 is the most preferable one. This solution is equivalent to others with respect to criterion *P*4; it is superior to others over a set of five criteria {*P*2, *P*3, *P*6, *FQ*1, *FQ*2} and is weaker than the prototype only with respect to the criterion P5. Furthermore, the remaining 19 solutions are better than the prototype with respect to four criteria {*P*2, *P*3, *FQ*1, *FQ*2}. However, none of the solutions are more superior than the prototype with respect to criterion P5. In particular, this observation implies that if the prototype design vector was sampled by the system, then it would belong to the Pareto set.

The analysis of the histograms demonstrates the filtering effect of the criteria constraints (see Figure 9.3). The feasible solutions for the design variables *DV*1 and *DV*2 are located in the middle of the interval, with no feasible solutions in the left and right ends of the intervals, and slightly shifted from the prototype design (marked with a yellow triangle ∇). These histograms clearly identify feasible intervals for the design variables *DV*1 and *DV*2 characterizing

Table 9.2
Criteria Constraints

$P2 \leq 0.2$	(min)	$P9 \leq 15$	(pseudo)
$P3 \leq 0.2$	(min)	$P10 \leq 300$	(pseudo)
$P4 \leq 1.02$	(min)	$P11 \leq 0.25$	(pseudo)
$P5 \leq 1$	(min)	$P12 \leq 0.01$	(pseudo)
$P6 \leq 0.3$	(min)	$P13 \leq 0.01$	(pseudo)
$P7 \leq 0.25$	(pseudo)	$FQ1 \leq 0.1$	(min)
$P8 \leq 5$	(pseudo)	$FQ2 \leq 45$	(min)

Table 9.3
Fragment of Criteria Table

Criteria		Prototype	Pareto Optimal Solutions						
			#241	**#281**	**#329**	**#409**	**#649**	**#825**	**#993**
*P*2	(min)	1.30E-01	1.04E-01	8.40E-02	9.14E-02	8.97E-02	6.03E-02	7.44E-02	5.39E-02
*P*3	(min)	1.54E-01	1.16E-01	1.03E-01	1.06E-01	1.04E-01	8.72E-02	9.51E-02	8.84E-02
*P*4	(min)	1.0E+00	1.00E+00	1.00E+00	1.00E+00	1.00E+00	1.01E+00	1.00E+00	1.00E+00
*P*5	(min)	3.15E-01	5.37E-01	9.36E-01	9.68E-01	6.29E-01	8.63E-01	8.58E-01	6.89E-01
*P*6	(min)	1.49E-01	1.97E-01	2.21E-01	2.58E-01	2.03E-01	1.74E-01	1.94E-01	1.32E-01
*P*7	(pseudo)	1.51E-01	1.65E-01	1.74E-01	1.84E-01	1.72E-01	1.83E-01	1.78E-01	1.75E-01
*P*8	(pseudo)	3.24E+00	3.3E+00	3.29E+00	3.31E+00	3.30E+00	3.31E+00	3.30E+00	3.31E+00
*P*9	(pseudo)	5.96E+00	1.17E+01	7.55E+00	9.1E+00	1.09E+01	1.11E+01	9.36E+00	1.16E+01
*P*10	(pseudo)	1.07E+02	2.42E+02	1.34E+02	1.72E+02	2.16E+02	2.24E+02	1.76E+02	2.44E+02
*P*11	(pseudo)	7.45E-02	7.79E-02	6.02E-02	6.82E-02	8.10E-02	7.72E-02	7.22E-02	7.77E-02
*P*12	(pseudo)	1.01E-04	1.84E-04	1.87E-04	2.08E-04	2.06E-04	2.29E-04	2.14E-04	2.09E-04
*P*13	(pseudo)	3.16E-05	6.18E-05	6.87E-05	8.01E-05	7.00E-05	8.18E-05	7.58E-05	7.03E-05
*FQ*1	(min)	1.23E-02	6.73E-02	9.29E-02	9.74E-02	6.86E-02	9.05E-02	7.80E-02	8.85E-02
*FQ*2	(min)	5.36E+01	4.42E+01	4.35E+01	4.03E+01	4.22E+01	3.79E+01	4.10E+01	4.00E+01

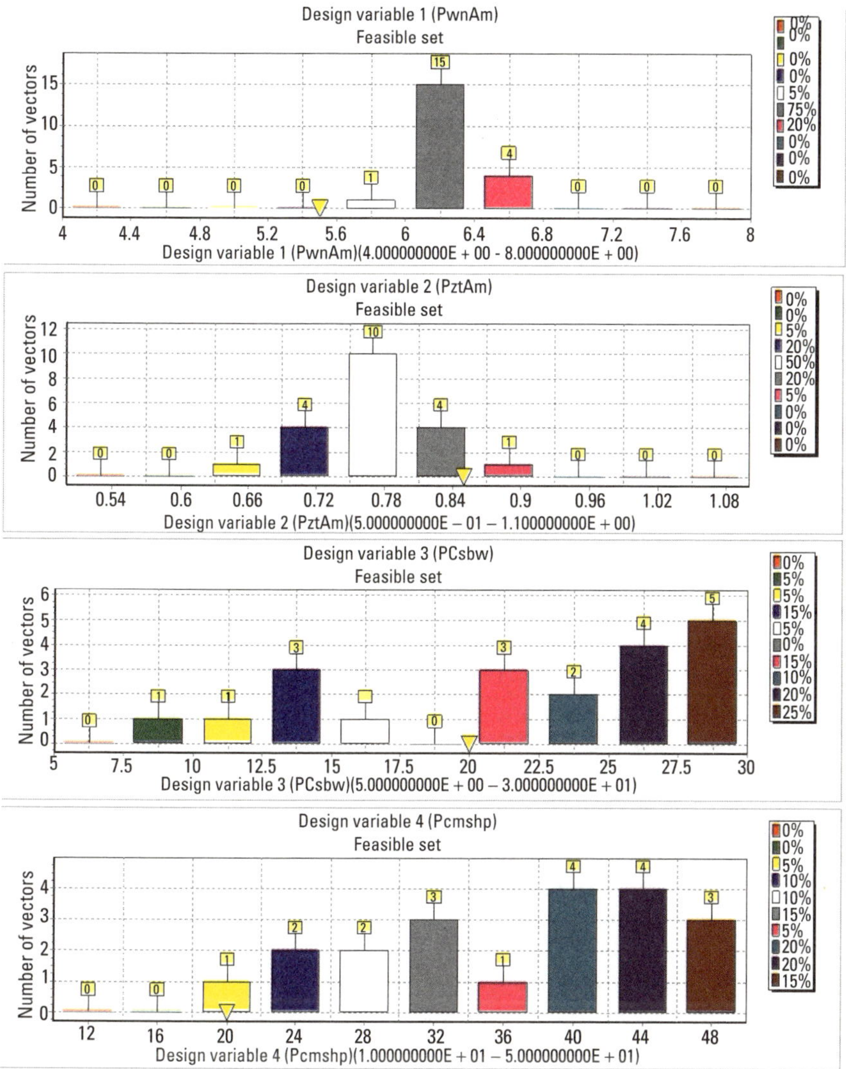

Figure 9.3 PSI iteration 1, distribution of feasible solutions; yellow triangle shows the prototype design.

the state-predictor dynamics. On the other hand, the feasible solutions for the design variables *DV*3 and *DV*4 are located at the right boundaries of the intervals. These observations will be used in the next step of the optimization process to modify the initial intervals in the direction of higher density of feasible solutions to improve richness of the feasible solution set.

It is also important to analyze the influence of the design variables on criteria and pseudo-criteria and to evaluate the degree of improvement (or degradation) of the Pareto optimal solutions with respect to the prototype design. Figures 9.4 to 9.8 show some of these dependencies, which provide valuable insight into the effect of the design variables on the augmented aircraft dynamics. Preliminary conclusions that can be drawn from this analysis are detailed next.

Figure 9.4 shows a typical dependency of criteria *P2* (*P3* and *P6* have the same trend) on the design variable *DV1*. First, we notice that there are no feasible solutions in the range 4 < *DV1* < 4.6, which indicates that such design variable vectors did not satisfy the constraints. Second, it can be seen that the deviations of the actual response (*P2*, *P3*, and *P6*) of the augmented aircraft from the desired model become large for big values of the design variable *DV1*. Third, all three dependencies show that the set of Pareto optimal solutions, when compared to the prototype design, provide step responses with reduced deviations in terms of both (truncated) L_∞ and L_2 norms. The rate deviations for the Pareto optimal solutions are, however, larger than the rate deviations for the prototype design.

Figure 9.5 presents the dependencies of the flying qualities criteria *FQ2* (*FQ1* has the same trend) on the design variable *DV1*, which indicate that faster responses of the augmented aircraft (large *DV1*) result in improved (predicted) flying qualities. This trend is particularly evident in Figure 9.5, in which the reduced pilot (lead) compensation is limited to large values of the design variable *DV1*. Also, when compared to the prototype design, the set of optimal solutions improve the criterion *FQ1* by 20% to 50%, and the criterion *FQ2* by 15% to 30%.

As expected, a faster response of the augmented aircraft (large *DV1*) leads to an increase in the control effort (pseudo-criteria P8), as well as an increase in both the maximum elevator rate and the maximum elevator acceleration (pseudo-criteria *P9* and *P10*).

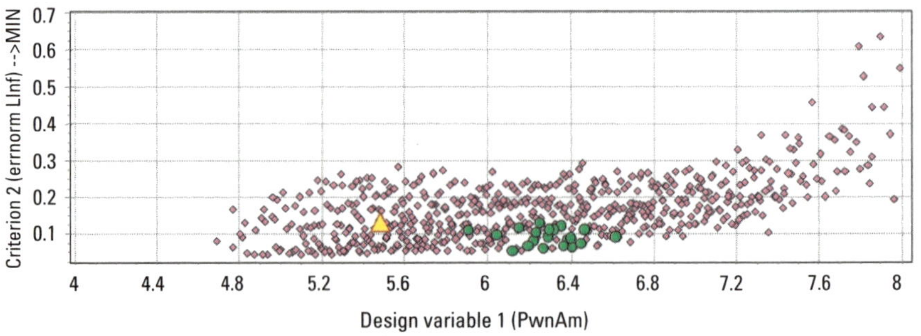

Figure 9.4 PSI iteration 1; dependencies of the criterion *P2* (maximum AOA deviation) on the design variable *DV1*.

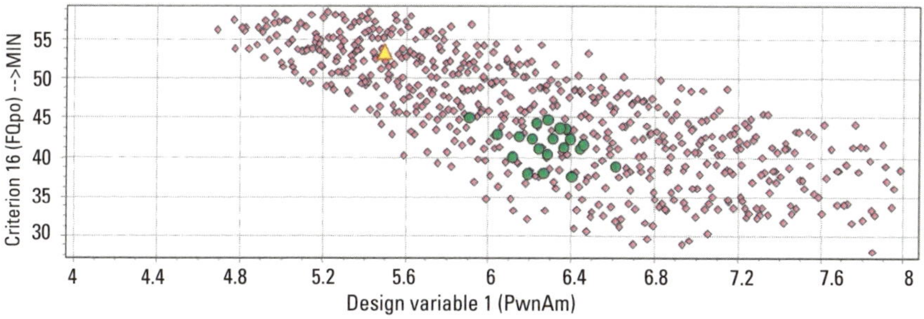

Figure 9.5 PSI iteration 1; dependencies of the criterion *FQ2* (pilot workload) on the design variable *DV1*.

Figure 9.6 illustrates the dependencies of criteria *FQ*1 on the design variable *DV*2. (*FQ*2 has the same trend.) While the dependency of *FQ*2 on *DV*2 is not obvious, small values of *DV*2 seem to limit the achievable tracking performance with the pilot in the loop (*FQ*1). This would imply that the augmented aircraft with low-damping characteristics result in degraded (predicted) flying qualities.

Finally, the dependencies of pseudo-criteria *P*9 and *P*10 on the design variable *DV*4 show that the Pareto optimal solutions are located along a "straight" line with positive slope where high values of the criteria correspond to high values of the design variable *DV*4. This implies that the bandwidth of the command prefilter in the L_1 FCS (*DV*4) can be set to limit the maximum elevator rate and the maximum elevator acceleration of the Pareto optimal solutions.

Dependencies between criteria provide useful information about the solutions in the Pareto set and the trade-offs between criteria. The analysis of trade-offs becomes especially helpful in the final stages of the optimization process, when a decision about the most preferable solution has to be made. These

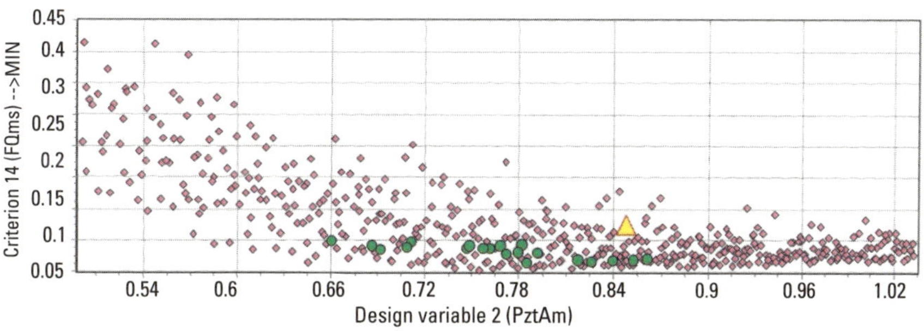

Figure 9.6 PSI iteration 1; dependencies of the criterion *FQ1* (tracking performance) on the design variable *DV2*.

dependencies are useful to explore the trade-offs in the design of the controller and to identify directions in the design variable space that may lead to improved L_1 FCS design over the prototype design.

Next we present and discuss a set of results showing dependencies between criteria illustrating these trade-offs and determining possible directions of improvement of the prototype solution:

Figure 9.7 shows the dependency between *P*2 and *P*3 (maximum and integral deviations from the desired response) as well as the localization of the Pareto optimal solutions with respect to the prototype design in the *P*2-*P*3 plane. It can be seen that the solutions of the Pareto set are located on a "straight" line passing through the prototype design, and all of them improve the prototype in terms of the metrics *P*2 and *P*3.

Figure 9.8 shows the location of the prototype as well as the Pareto optimal solutions in the FQ plane of the TDNS criteria, thus characterizing the (predicted) flying qualities of the solutions. The improvement with respect to the prototype design in terms of predicted FQ is evident in this *FQ*2-*FQ*1

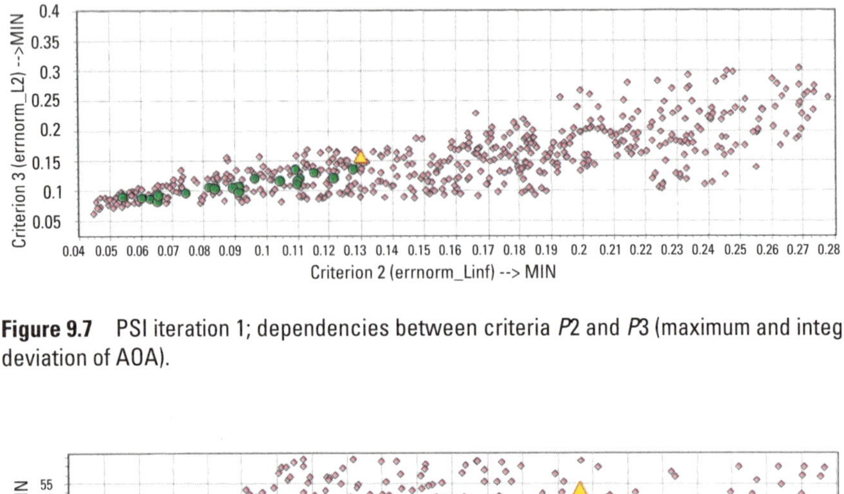

Figure 9.7 PSI iteration 1; dependencies between criteria *P*2 and *P*3 (maximum and integral deviation of AOA).

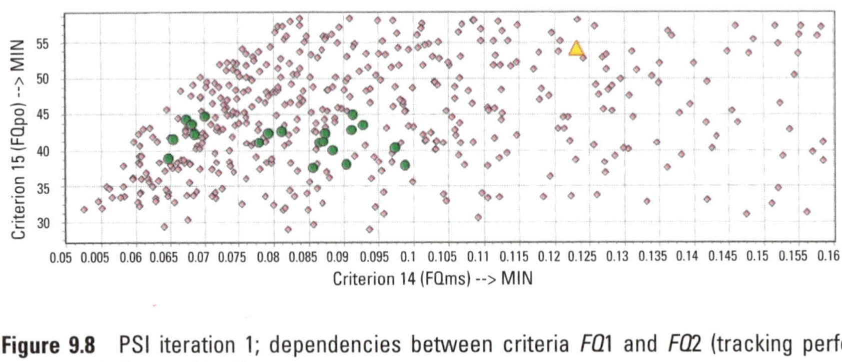

Figure 9.8 PSI iteration 1; dependencies between criteria *FQ*1 and *FQ*2 (tracking performance and pilot workload).

graph. Also, the boundary for level 1 FQ of the TDNS criterion can be easily recognized in this figure (see [16]).

Finally, Figure 9.9 shows the dependency between the FQ criterion *FQ2* and the control effort *P8*. This figure shows that reduced pilot (lead) compensation is only possible for increased control effort. In fact, all of the solutions in the Pareto set present higher control efforts than the prototype solution.

The analysis of the obtained solutions allows us to define the direction of further search. In particular, the results have provided tight intervals for the design variables *DV1* and *DV2* characterizing the state-predictor dynamics, and have exposed the necessity of extending the initial intervals of initial intervals of variation of the design variables *DV3* and *DV4*. It is worth emphasizing that this significant improvement of the prototype solution has been achieved utilizing a reduced number of criteria, {*P2*, *P3*, *P4*, *P5*, *P6*, *FQ1*, *FQ2*}. Based on these results and the conclusions drawn from them, a new experiment is carried out to: (1) improve the feasible solution set, and (2) determine an optimal solution of the L_1 FCS design that improves the prototype with respect to extended set of criteria {*P2*, *P3*, *P4*, *P5*, *P6*, *FQ1*, *FQ2*, *FQ4*, *R1*}.

9.3.2 Second Iteration

The improvement of the feasible solution set is based on the analysis of the histograms in Figure 9.3 and Table 9.3. The analysis of the histograms results in adjusting the initial problem statement by changing the intervals of variation of the design variables and is given in Table 9.4.

The criteria constraints remain unchanged, whereas the design is now to be optimized with respect to the following new set of criteria {*P2*, *P3*, *P4*, *P5*, *P6*, *FQ1*, *FQ2*, *FQ4*, *R1*}. All these criteria are to be minimized except for *R1*, which is to be maximized.

Next we present the results obtained in this second iteration of the optimization process, which are based on 512 tests. As mentioned previously, adding

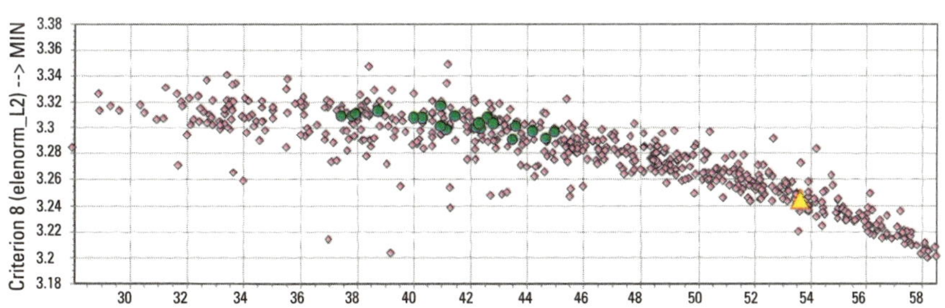

Figure 9.9 PSI iteration 1, dependencies between criteria *FQ2* and *P8* (pilot workload and maximum AOA deviation).

Table 9.4
Refined Intervals of Design Variables

Design Variable	Prototype	Initial Intervals of Variation of Design Variables	
		Min	**Max**
*DV*1	5.50E+00	5.50E+00	7.00E+00
*DV*2	8.50E-01	6.50E-01	0.90E+00
*DV*3	2.00E+01	9.80E+00	4.00E+01
*DV*4	2.00E+01	1.80E+01	6.50E+01

criteria *R*1 and *FQ*4 in the optimization process results in a significant increase in the computational time. Approximately 124 vectors satisfied previously assigned criteria constraints (see Table 9.2). All these solutions are Pareto optimal. As might be expected, the coefficient of selection efficiency γ increased by more than 12 times. The histograms in the second iteration have better distributions (higher concentration) of the feasible solutions than in the first iteration. The top of Figure 9.10 represents the distribution of 124 solutions for *DV*1.

After the analysis of the test table, stronger criteria and pseudo-criteria constraints were considered as presented in Table 9.5. According to these new constraints, only six solutions were found to be feasible, and all of them are

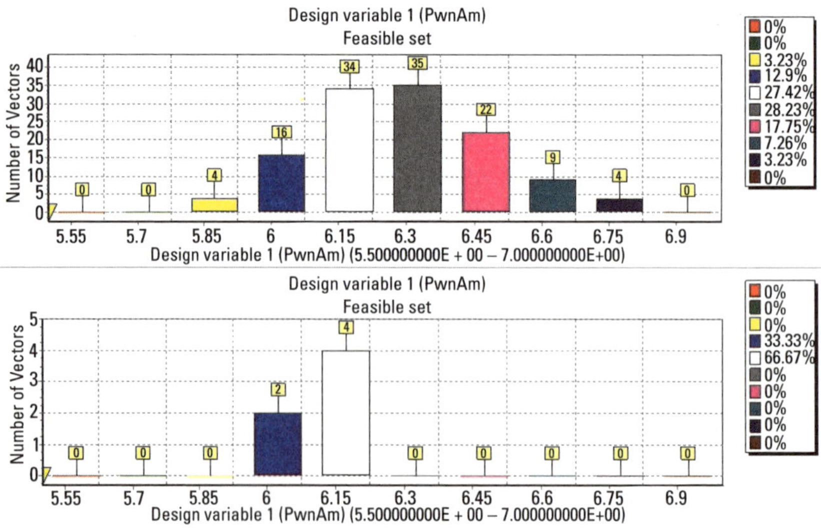

Figure 9.10 PSI iteration 2, distribution of feasible solutions of *DV*1 with the original (top) and with tightened criteria constraints (bottom).

Table 9.5
Second Iteration, Refined Criteria Constraints

$P2 \le 0.1$	(min)	$P10 \le 200$	(pseudo)
$P3 \le 0.15$	(min)	$P11 \le 0.1$	(pseudo)
$P4 \le 1.02$	(min)	$P12 \le 0.01$	(pseudo)
$P5 \le 1$	(min)	$P13 \le 0.01$	(pseudo)
$P6 \le 0.25$	(min)	$FQ1 \le 0.1$	(min)
$P7 \le 0.2$	(pseudo)	$FQ2 \le 45$	(min)
$P8 \le 5$	(pseudo)	$FQ4 \le 5$	(min)
$P9 \le 10$	(pseudo)	$R1 \ge 80$	(max)

Pareto optimal. The values of design variables and criteria of the Pareto optimal solutions are given in Tables 9.6 and 9.7. The new distribution of the feasible solutions for these criteria and pseudo-criteria constraints is significantly tighter as expected and is shown in the bottom of Figure 9.10. These new histograms clearly identify tight intervals for all of the design variables in which the optimal solutions lie.

Furthermore, from the analysis of Table 9.6, it follows that all solutions of second iteration as well as the #993 from the first iteration belong to the very tight intervals of the first and second design variables. The first three parameters (*DV1*–*DV3*) of #993 and #106 are almost identical. Observe that #993, while providing good response of many criteria, does not satisfy new constraints on criteria *P9* and *P10* (the elevator workload). Moreover, the analysis of the #993 also shows that it fails to satisfy constraint of the flying qualities criteria *FQ3*.

The analysis of test tables, dependencies of criteria on design variables, and dependencies between criteria allows us to determine the most preferable solutions. In particular, Figure 9.11 shows the influence of the bandwidth of the "matched" lowpass filter (*DV3*) on the (pilot-off-the-loop) trade-off between performance criterion *P2* (*P3* shows the same trend) and robustness (*R*1) of the augmented aircraft. From this observation we can conclude that criteria *P2* (*P3*) and *R*1 are contradictory with respect to the design variable *DV3*. This means that improvement of the tracking performance requires an increase in the bandwidth of the lowpass filter, which in turn results in degradation of the time delay margin of the augmented aircraft, as predicted by theory.

Figure 9.12 shows the dependencies of the flying qualities criterion *FQ*1 (*FQ2* shows a similar trend) on the design variable *DV2*. While in the first PSI iteration the dependency of the criterion *FQ2* on the design variable *DV2* was not obvious, now it becomes more apparent that a smaller damping ratio seems to result in reduced (lead) pilot compensation.

Table 9.6
Second Iteration: Table of Design Variables

Design variable	Prototype	#993, First Iteration	Pareto Optimal Solutions					
			#106	#202	#254	#318	#358	#462
DV1	5.50E+00	6.12E+00	6.00E+00	5.99E+00	6.24E+00	6.23E+00	6.10E+00	6.18E+00
DV2	8.50E+00	7.09E-01	7.34E-01	7.49E-01	7.76E-01	7.33E-01	7.81E-01	7.18E-01
DV3	2.0E+01	2.70E+01	2.52E+01	1.67E+01	1.81E+01	2.18E+01	1.69E+01	1.58E+01
DV4	2.0E+01	4.93E+01	3.16E+01	3.20E+01	2.10E+01	2.72E+01	2.57E+01	3.11E+01

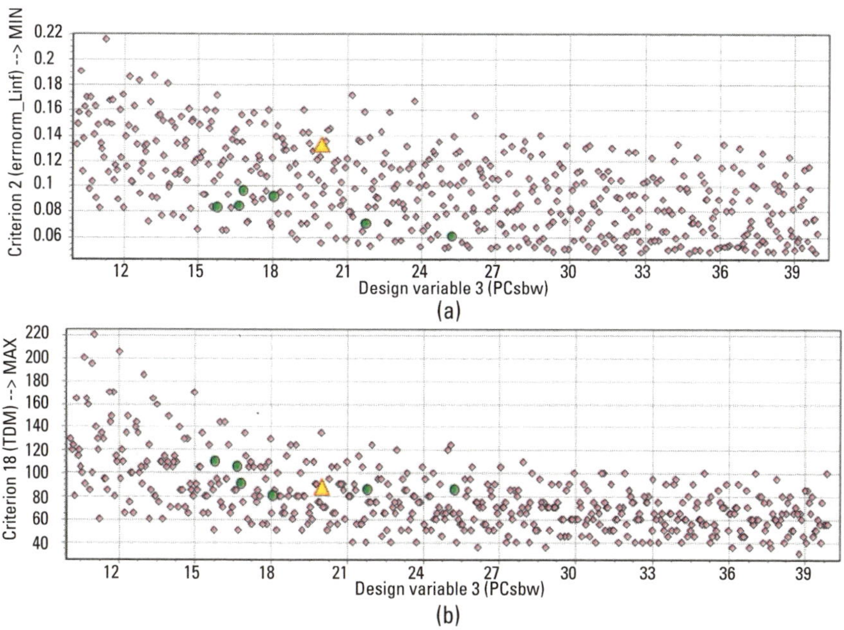

Figure 9.11 PSI iteration 2, dependencies of criteria *P2* and *R1* on the design variable *DV3*. (a) Criterion *P2* (max AOA deviation) versus design variable *DV3*. (b) Criterion *R1* (time delay margin) versus design variable *DV3*.

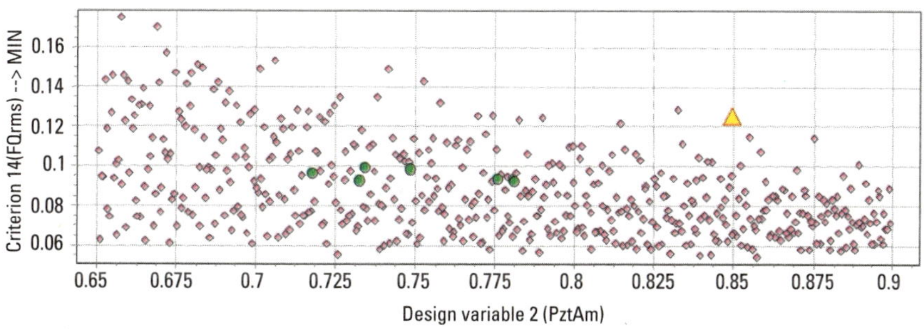

Figure 9.12 PSI iteration 2, dependencies of criteria *FQ1* (tracking performance) on the design variable *DV2*.

Dependency between *P2* and *P3* is analogous to those obtained in first iteration; therefore, the corresponding figure is not presented. All of the Pareto optimal solutions improve the prototype design in terms of peak and integral deviations from the desired response.

The dependency of flying qualities criteria *FQ1* and *FQ2* obtained in this iteration is also similar to those obtained in the first iteration, thus demonstrating

Table 9.7
Second Iteration: Table of Criteria

Criteria		Prototype	Pareto Optimal Solutions #106	#202	#254	#318	#358	#462
P2	(min)	1.30E-001	6.01E-01	8.45E-02	9.17E-02	7.04E-02	9.63E-02	8.28E-02
P3	(min)	1.54E-01	9.60E-02	1.13E-01	1.11E-01	9.66E-02	1.18E-01	1.04E-01
P4	(min)	1.0E+00	1.00E+00	1.00E+00	1.00E+00	1.00E+00	1.01E+00	1.00E+00
P5	(min)	3.15E-01	6.13E-01	6.81E-01	8.85E-01	8.43E-01	7.71E-01	8.63E-01
P6	(min)	1.49E-01	1.28E-01	1.67E-01	2.16E-01	1.78E-01	1.94E-01	2.11E-01
P7	(pseudo)	1.51E-01	1.68E-01	1.67E-01	1.72E-01	1.76E-01	1.68E-01	1.78E-01
P8	(pseudo)	3.24E+00	3.29E+00	3.29E+00	3.29E+00	3.30E+00	3.29E+00	3.31E+00
P9	(pseudo)	5.96E+00	9.09E+00	9.1E+00	7.9E+00	9.05E+00	8.43E+00	9.54E+00
P10	(pseudo)	1.07E+02	1.77E+02	1.78E+02	1.43E+02	1.72E+02	1.59E+02	1.85E+02
P11	(pseudo)	7.45E-02	6.62E-02	6.60E-02	6.09E-02	6.74E-02	6.31E-02	6.95E-02
P12	(pseudo)	1.01E-04	1.81E-04	1.70E-04	1.80E-04	2.01E-04	1.69E-04	1.98E-04
P13	(pseudo)	3.16E-05	6.02E-05	5.89E-05	6.58E-05	7.13E-05	6.01E-05	7.27E-05
FQ1	(min)	1.23E-02	9.93E-02	9.81E-02	9.34E-02	9.26E-02	9.20E-02	9.64E-02
FQ2	(min)	5.36E+01	4.33E+01	4.41E+01	4.38E+01	4.14E+01	4.46E+01	4.10E+01
FQ4	(min)	4.68E+00	4.08E+00	4.19E+00	3.87E+00	3.88E+00	3.84E+00	3.97E+00
R1	(max)	8.50E+01	8.50E+01	1.05E+02	8.00E+01	8.50E+01	9.00E+01	1.00E+02

significant improvement of predicted flying qualities over the prototype design but now in the extended criteria space.

Finally, Figure 9.13 shows the dependency between criteria *P*3 and *R*1, which illustrates the trade-off between performance and robustness of the closed-loop adaptive system with the pilot-off-the-loop. While all of the optimal solutions reduce the deviations from the desired response with respect to the prototype design, only three of these solutions exhibit a better time-delay margin than the prototype design (#202, #462), and two exhibit a similar margin (#106, #318)[3].

As a result of the iterative two-step correction of initial constraints, the six Pareto optimal solutions have been found. Analyzing the criteria table shows that there are no solutions better than the prototype by all criteria simultaneously. This analysis revealed that four solutions, #106, #202, #358, and #462, surpassed the prototype by six criteria simultaneously.

Although all six solutions are practically equivalent, some advantage is given to the design vector #202 since it provides a better trade-off between the (predicted) flying qualities (*FQ*1, *FQ*2) and the time-delay margin (*R*1), while minimizing the difference with the desired response (see Figure 9.14). The comparative analysis of the AOA and the elevator workload demonstrates that the optimal solution provides faster response (rise time) to the commanded AOA command while minimally increasing the elevator workload that is naturally expected.

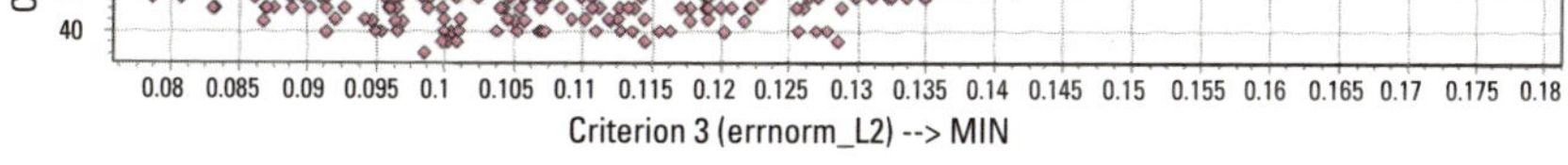

Figure 9.13 PSI iteration 2; dependencies between criteria *R*1 (time delay margin) and *P*3 (integral deviation of AOA).

3. Notice that for the determination of the time-delay margin, metric *R*1, we insert incremental time delays of 5 ms in the elevator command channel. This choice was made for consistency with the determination of the time-delay margin in the AirSTAR real-time simulator at NASA LARC. As a consequence, however, we cannot identify the time-delay margin with higher accuracy.

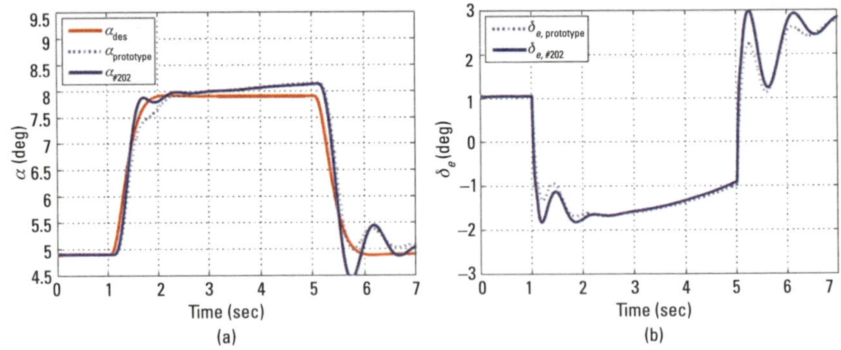

Figure 9.14 Optimal design #202, 3° AOA step response. (a) Angle of attack, α; (b) elevator deflection, δ_e.

9.3.3 Conclusion

The construction of the feasible solutions of the L_1 adaptive controller satisfying the desired performance and robustness specifications is a critical step for the optimal design of the L_1 FCS. In particular, the eigenstructure of the state-predictor state matrix and the bandwidth of the lowpass filters are the key elements that characterize the performance of the L_1 FCS. To optimize the design of these elements, a 16-criteria problem of improving a prototype solution was formulated and solved. The results presented demonstrate the application of PSI method to the multicriteria design optimization of the L_1 FCS implemented on the GTM AirSTAR aircraft. The study has addressed both the construction of the feasible solution set and the improvement of a nominal prototype design initially synthesized by the L_1 theory.

On one hand, the results have demonstrated that the consistent application of the systematic design guidelines of L_1 adaptive control becomes particularly beneficial for the construction of the feasible solution set. Moreover, the results of this study are consistent with the theoretical claims of the theory of L_1 adaptive control in terms of robustness and performance. On the other hand, the developed procedure and the obtained results confirm the suitability of the PSI method and the MOVI software package for the multicriteria optimization of an adaptive FCS subject to desired control specifications.

The optimal design of the L_1 FCS significantly improved understanding of the design trade-offs between adaptation and robustness. Finally, the design guidelines learned during the interaction with GTM model utilizing MOVI software contributed to the successful flight verification and validation of the designed all-adaptive control law at NASA LARC.

References

[1] Jacklin, S. A., et al., "Verification, Validation, and Certification Challenges for Adaptive Flight-Critical Control System Software," *AIAA Guidance, Navigation and Control Conference,* AIAA-2004-5258, Providence, RI, August 2004.

[2] Wise, K. A., E. Lavretsky, and N. Hovakimyan, "Adaptive Control in Flight: Theory, Application, and Open Problems," *American Control Conference,* Minneapolis, MN, June 2006, pp. 5966–5971.

[3] Jacklin, S. A., "Closing Certification Gaps in Adaptive Flight Control Software," *AIAA Guidance, Navigation and Control Conference,* AIAA-2008-6988, Honolulu, HI, August 2008.

[4] Hovakimyan, N., and C. Cao, *L_1 Adaptive Control Theory,* Philadelphia, PA: Society for Industrial and Applied Mathematics, 2010.

[5] Jordan, T. L., W. M. Langford, and J. S. Hill, "Airborne Subscale Transport Aircraft Research Testbed-Aircraft Model Development," *AIAA Guidance, Navigation and Control Conference,* AIAA-2005-6432, San Francisco, CA, August 2005.

[6] Jordan, T. L., et al., "AirSTAR: A UAV Platform for Flight Dynamics and Control System Testing," *AIAA Aerodynamic Measurement Technology and Ground Testing Conference,* AIAA-2006-3307, San Francisco, CA, June 2006.

[7] Xargay, E., N. Hovakimyan, and C. Cao, "L_1 Adaptive Controller for Multi-Input Multi-Output Systems in the Presence of Nonlinear Unmatched Uncertainties," *American Control Conference,* Baltimore, MD, June–July 2010.

[8] Gregory, I. M., et al., "L_1 Adaptive Control Design for NASA AirSTAR Flight Test Vehicle," *AIAA Guidance, Navigation and Control Conference, AIAA-2009-5738,* Chicago, IL, August 2009.

[9] Gregory, I. M., et al., "Flight Test of an L_1 Adaptive Controller on the NASA AirSTAR Flight Test Vehicle," *AIAA Guidance, Navigation and Control Conference,* Toronto, Canada, August 2010.

[10] Kim, K. K., and N. Hovakimyan, "Development of Verification and Validation Approaches for L_1 Adaptive Control: Multi-Criteria Optimization for Filter Design," *AIAA Guidance, Navigation and Control Conference,* Toronto, Canada, August 2010.

[11] Statnikov, R. B., and J. B. Matusov, *Multicriteria Analysis in Engineering,* Dordrecht/Boston/London: Kluwer Academic Publishers, 2002.

[12] Statnikov, R. B., et al., *MOVI 1.4 (Multicriteria Optimization and Vector Identification) Software Package,* Certificate of Registration, Registere of Copyright, U.S.A., Registration Number TXU 1-698-418, Date of Registration: May 22, 2010.

[13] Sobol, I. M., and R. B. Statnikov, *Selecting Optimal Parameters in Multicriteria Problems,* 2nd ed., (in Russian), Moscow: Drofa, 2006.

[14] Xargay, E., et al., "L_1 Adaptive Flight Control System: Systematic Design and V&V of Control Metrics," *AIAA Guidance, Navigation, and Control Conference,* Toronto, Ontario, Canada, August 2–5, 2010.

[15] Stepanyan, V., et al., "Stability and Performance Metrics for Adaptive Flight Control," *AIAA Guidance, Navigation and Control Conference,* Chicago, IL, AIAA-2009-5965, August 2009.

[16] Bailey, R. E. and T. J. Bidlack, *Unified Pilot-Induced Oscillation Theory, Volume IV: Time-Domain Neal-Smith Criterion,* Tech. Rep. WL-TR-96-3031, Air Force Wright Laboratory, December 1995.

[17] Choe, R., et al., "L_1 Adaptive Control Under Anomaly: Flying Qualities and Adverse Pilot Interaction," *AIAA Guidance, Navigation and Control Conference,* Toronto, Canada, August 2010.

Conclusions

All real-life optimization problems, such as the design, identification, design of controlled systems, and operational development of prototypes, are essentially multicriteria. Constraints on the design variables, functional dependences, and criteria determine the feasible solution set. This set is the domain where one should be looking for the optimal solutions.

A statement of real-life optimization problem primarily involves justification of the feasible set. It is usually assumed that the statement of the optimization problem is the prerogative and skill of the expert. However, there are few who understand that the expert is practically unable to define the feasible set in the best way. The skill of the expert is a necessary, but by no means sufficient condition for the correct statement of the real-life problem. Since criteria are contradictory, the definition of the feasible solution set represents significant, sometimes insurmountable, difficulties. As a result, the expert solves the incorrectly stated optimization problem. If the problem has not been stated correctly, the application of optimization methods, however good they may be, is often ineffective. This situation is common today when solving real-life problems. Unfortunately, the expert will not be able to find the instructions on how to correctly determine the feasible solution set, and hence to state the optimization problem. All the numerous works devoted to real-life optimization discuss how to solve the problem but not how to state it correctly.

In order to construct the feasible solution set, a method called the parameter space investigation (PSI) has been created and successfully integrated into various fields of industry, science, and technology.

On the basis of the PSI method and MOVI software, we have: (1) demonstrated how to help an expert state and solve optimization problems and thus answered the fundamental question of where to search for optimal solutions; (2) explained the necessity of multicriteria analysis for constructing a feasible solution set; (3) shown tools for multicriteria analysis; and (4) presented the statement and solution of real-life optimization problems.

The correct determination of the feasible solution set is a major challenge in real-life optimization problems. Multicriteria analysis of the feasible solution set always provides the valuable information for decision-making. This analysis demonstrates the work of all constraints; the cost of making concessions in various constraints (i.e., what are the losses and the gains, expediency of modification of constraints, and resources for improving the object in all criteria).

The definition of the feasible solution set on the basis of the PSI method/ MOVI software has, for the first time, allowed the correct statement and solution of real-life problems.

Appendix
Examples of Calculation of the Approximate Compromise Curves

This appendix considers two simple problems where compromise curves can be found by analytical procedure; the calculated approximate compromise curves are compared to the exact ones [1].

Analytical Approach

If two functions $\Phi_1(\alpha)$ and $\Phi_2(\alpha)$ are differentiable, then one can try to find the locus of contact of level surfaces $\Phi_1(\alpha) = b_1$ and $\Phi_2(\alpha) = b_2$, see [1, 2]. In these points

$$\mathbf{grad}\ \mathbf{\Phi}_1 = -\lambda\ \mathbf{grad}\ \mathbf{\Phi}_2$$

This vector equation is equivalent to a set of n scalar algebraic equations

$$\partial\Phi_1/\partial a_j = -\lambda\partial\Phi_2/\partial\alpha_j \quad (j = 1, 2, \ldots, n)$$

which, generally speaking, defines the following curve in the design variable space:

$$\alpha_1 = \varphi_1(\lambda), \ldots, \alpha_n = \varphi_n(\lambda)$$

If a segment of this curve, on which $\lambda \geq 0$, belongs to the feasible set D_α, then it belongs to the Pareto optimal set E_α as well. For example, Figure A.1, which corresponds to Example A.1, shows arc AA', which consists of Pareto optimal points, and arc $A'A''$, which consists of points of contact, which are not Pareto optimal, although both arcs are described by the same equation:

$$\alpha_2 = 4\alpha_1/(3\alpha_1 - 1)$$

In this case the segment of the compromise curve is determined by the following parametric equations:

$$\Phi_1 = \Phi_1(\varphi_1(\lambda),\ldots,\varphi_n(\lambda)),$$
$$\Phi_2 = \Phi_2(\varphi_2(\lambda),\ldots,\varphi_n(\lambda)), \quad (\lambda \geq 0)$$

Along this curve

$$d\Phi_1 = \sum_{j=1}^{n}(\partial\Phi_1/\partial a_j)d\varphi_j = -\lambda\sum_{j=1}^{n}(\partial\Phi_2/\partial a_j)d\varphi_j = -\lambda d\Phi_2$$

which gives the following expression for the inclination of the compromise curve:

$$d\Phi_2/d\Phi_1 = -1/\lambda$$

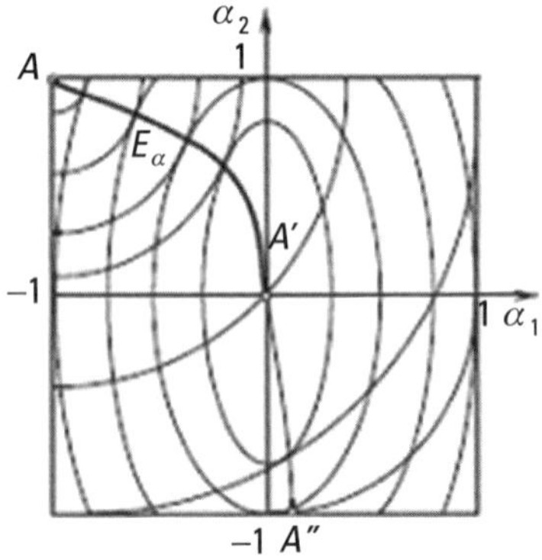

Figure A.1 Level curves Φ_1 = const and Φ_2 = const, set of points of contact $AA'A''$, Pareto optimal points—arc AA'.

Example A.1

The following two criteria are defined over square

$$D_\alpha = \{-1 \le \alpha_1 \le 1, -1 \le \alpha_2 \le 1\}:$$
$$\Phi_1 = 4\alpha_1^2 + \alpha_2^2 \text{ and } \Phi_2 = (\alpha_1 + 1)^2 + (\alpha_2 - 1)^2$$

these criteria need to be minimized.

Absolute minimums of the functions Φ_1 and Φ_2 are reached, respectively, at points (0, 0) and (–1, 1), which belong to D_α. Therefore, line E_α should link these points up. The above-mentioned analytical procedure gives a segment AA' of the hyperbola $a_2 = 4\alpha_1 (3\alpha_1 - 1)^{-1}$ shown in Figure A.2(a).

Part of D_α, located to the right of the line $AA'A''$, maps to the curvilinear figure $BB'B''B'''B_R$ at the criteria plane, while part of D_α located to the left of line $AA'A''$ maps to $BB'B''B_L$. Therefore, points located in the middle of D_Φ have two prototypes in D_α. Line BB' is the compromise curve [Figure A.2(b)].

The results of this problem's calculation are shown in Figures A.3 and A.4. Figure A.3 shows the exact compromise curve $\Phi_2 = u(\Phi_1)$ and the approximate line $\Phi_2 = u_N(\Phi_1)$. For $N = 511$, exact and approximate curves coincide within 0.01. Figure A.4 shows all trial points in D_α and in D_Φ. Notice the increased density of these points in the middle part of D_Φ.

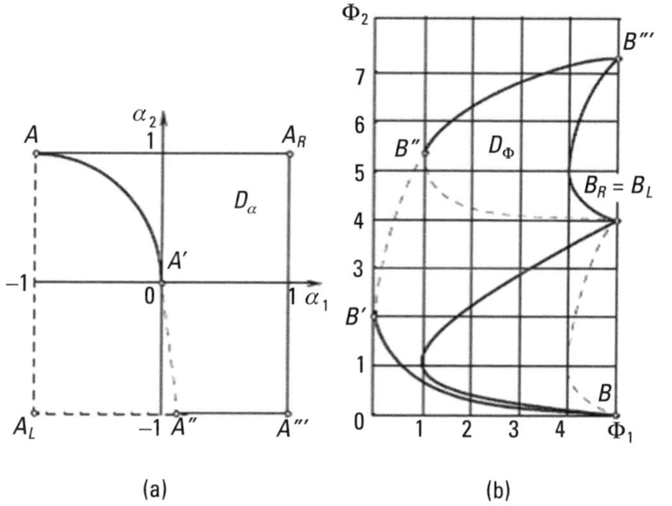

Figure A.2 (a) Feasible solution set D_α in the design variable space and (b) feasible solution set D_Φ in the criteria space. Arc AA' represents set $E_{\alpha'}$, while arc BB' represents set E_Φ.

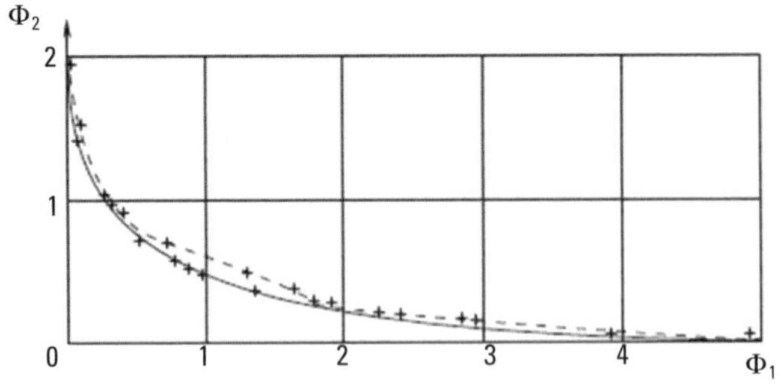

Figure A.3 Exact (solid line) and approximate (dotted line) compromise curve for $N = 63$. Crosses represent approximately Pareto optimal points for $N = 255$.

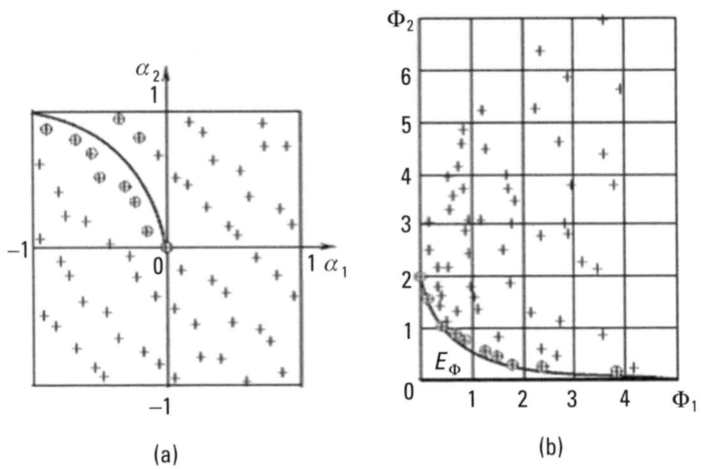

Figure A.4 Trial points (a) in D_α in design variable space and (b) in D_Φ in criteria space for $N = 63$. Approximate Pareto optimal points are circled.

Example A.2

The following two criteria are defined over square $\Pi = (-0.5 \le \alpha_1 \le 0.5,\ 0 \le \alpha_2 \le 1)$:

$$\Phi_1 = \alpha_1^2 + 4\alpha_2^2 \text{ and } \Phi_2 = (\alpha_1 + 1)^2 + (\alpha_2 - 1)^2$$

these criteria need to be minimized taking into account functional constraint

$$|\alpha_2 - \alpha_1 - 0.375| \ge 0.125$$

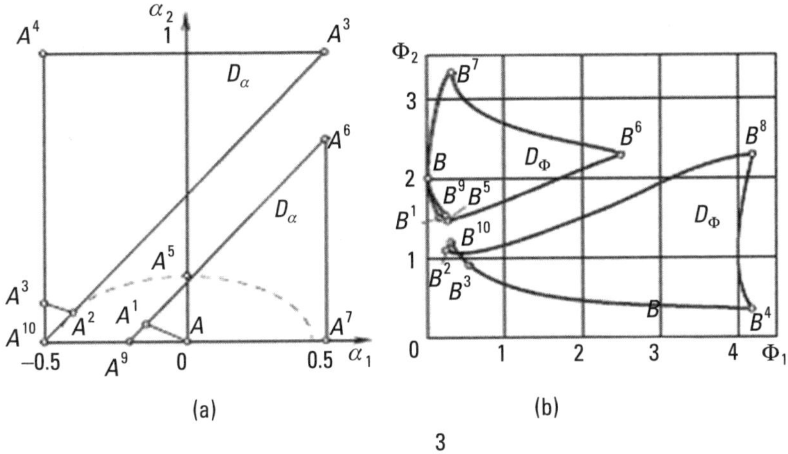

Figure A.5 Feasible solutions set (a) D_α in design variable space and (b) D_Φ in criteria space. Set E_α consists of two segments AA^1A^5 and $A^2A^3A^4$, while set E_Φ consists of segments BB^1B^5 and $B^2B^3B^4$.

In this problem set D_α is represented by square Π with a strip-shaped cutout (Figure A.5). Set E_α consists of two segments AA^1 and A^2A^3 of the hyperbola $\alpha_2 = -\alpha_1(3\alpha_1 + 4)^{-1}$, segment A^3A^4 of the border line $\alpha_1 = -0.5$, and segment A^1A^5 of the border line $\alpha_2 - \alpha_1 = 0.25$. Figure A.5 also shows set D_Φ on the criteria plane. However, those parts of D_Φ that consist of points that have two prototypes in D_α are much smaller here than in Example A.1.

The exact compromise curve is also shown in Figure A.6. Figure A.7 shows trial points in D_Φ. Results obtained for $N = 63$ are not sufficient to conclude that set D_Φ consists of two separate parts; however, results for $N = 255$ make that certain. Here q is number of points that satisfy the functional constraint $|\alpha_2 - \alpha_1 - 0.375| \geq 0.125$.

The computation above was performed using the LP_τ generator. Below we also show Example A.2 for 64, 256, 512, and 1,024 trials[1] (Figures A.8 through A.11) performed using a random number generator. In these figures the Pareto optimal set is marked in red in the criteria space.

1. This example has been performed by Douglas S. Parten.

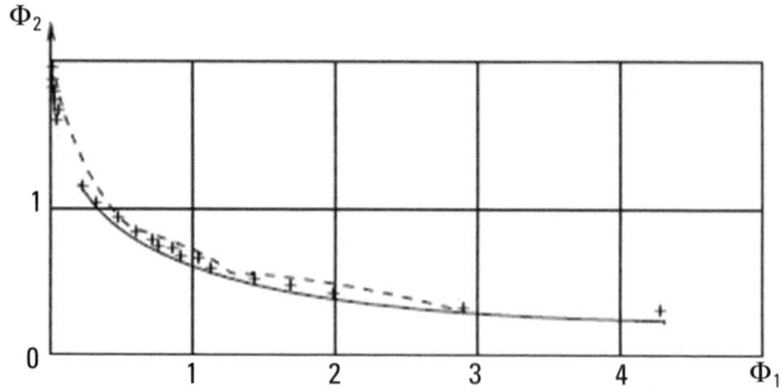

Figure A.6 Exact (solid line) and approximate (dotted line) compromise curves for $N = 63$. Crosses represent approximate Pareto optimal points for $N = 255$.

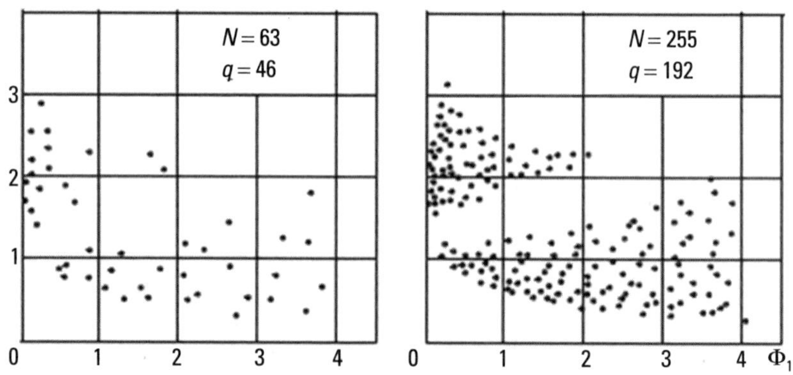

Figure A.7 All q trial points in the criteria space for $N = 63$ and for $N = 255$.

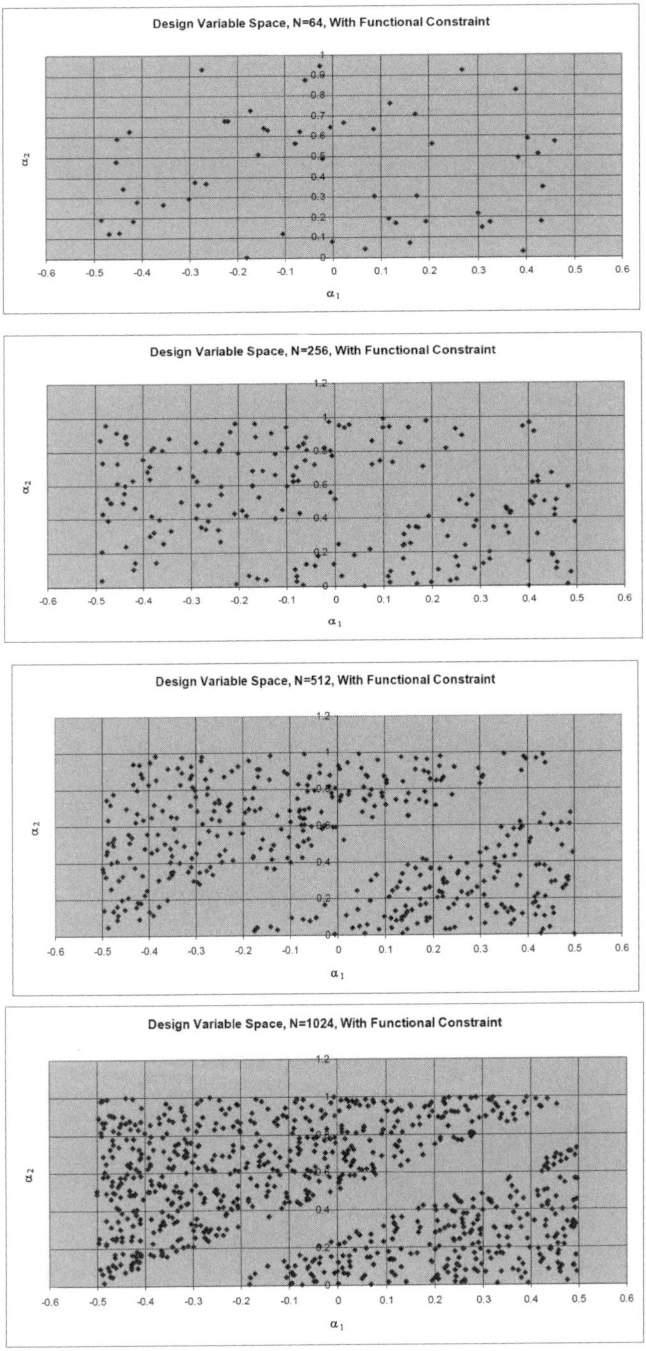

Figure A.8 Investigations with a functional constraint. The left column shows the trial points in the design variable space. The right column shows the trial points in the criteria space.

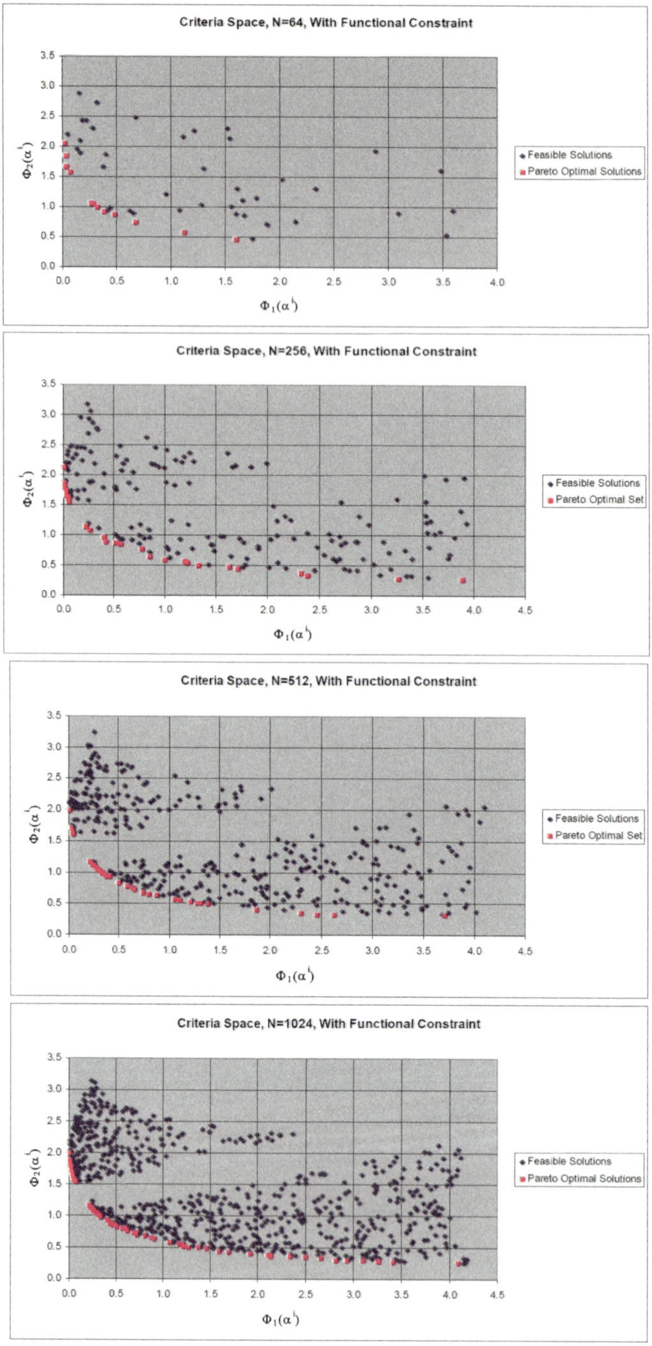

Figure A.8 (continued).

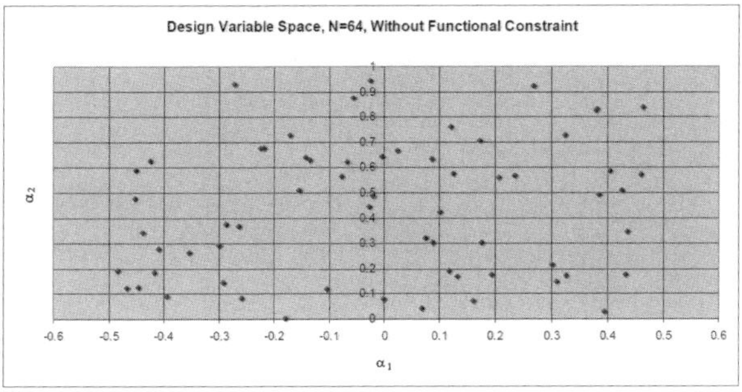

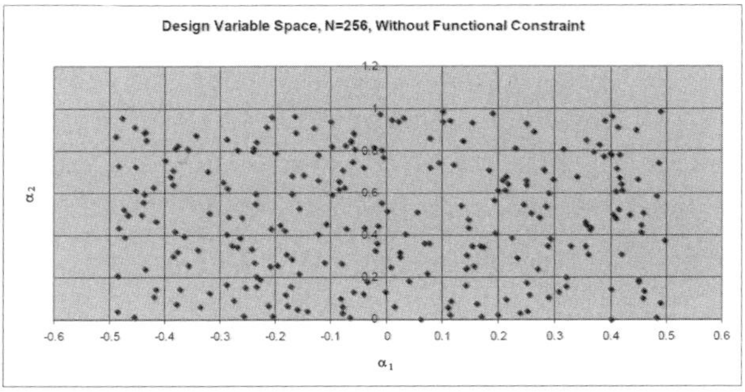

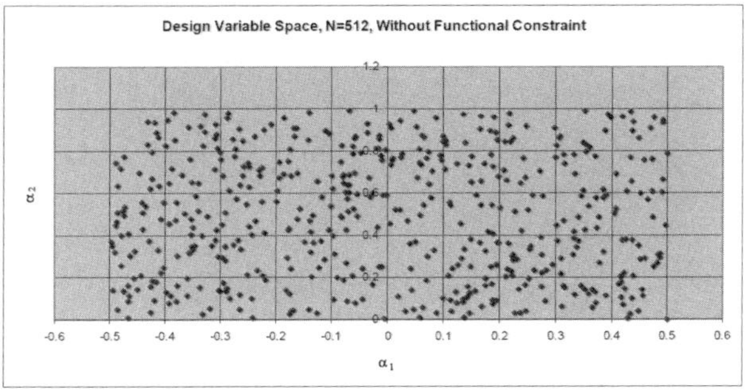

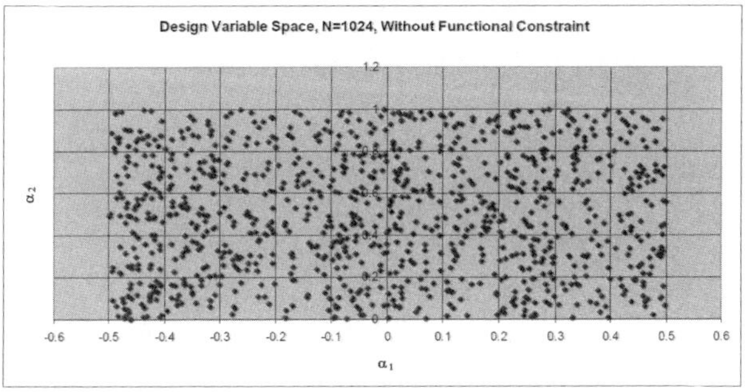

Figure A.9 Investigations without a functional constraint. The left column shows the trial points in the design variable space. The right column shows the trial points in the criteria space.

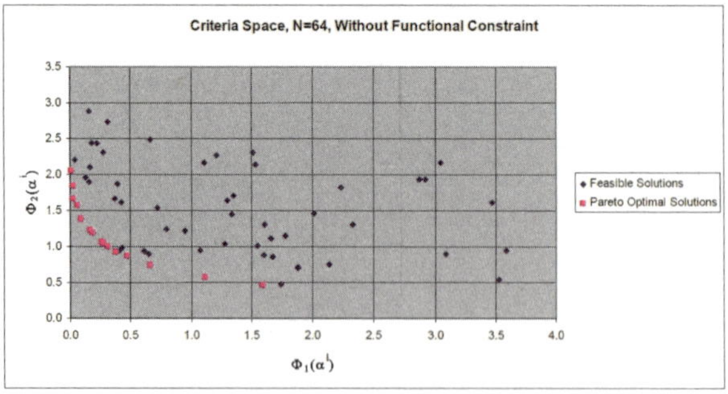

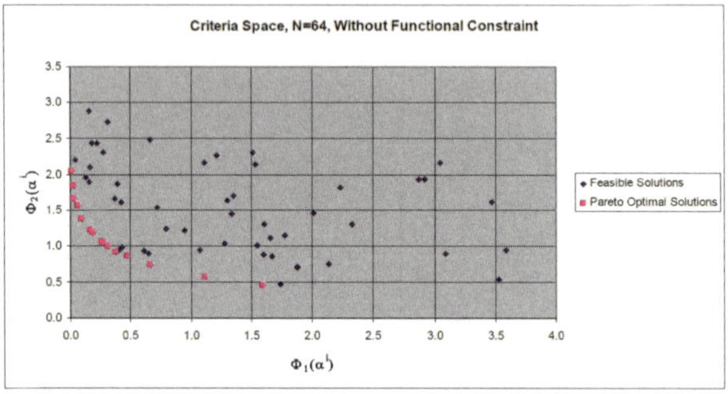

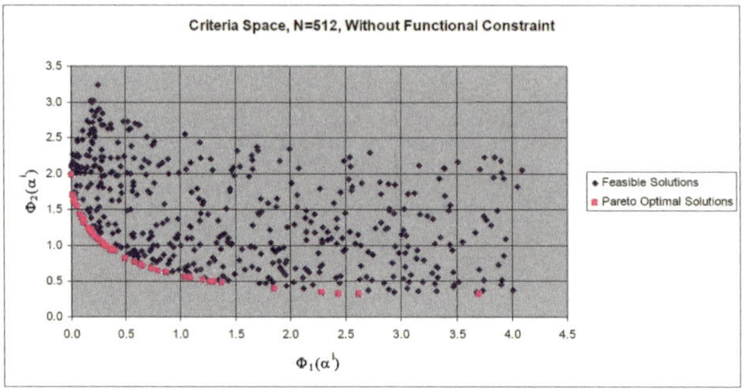

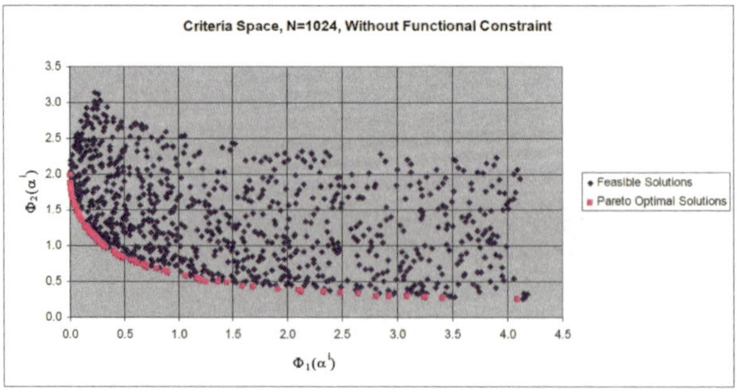

Figure A.9 (continued).

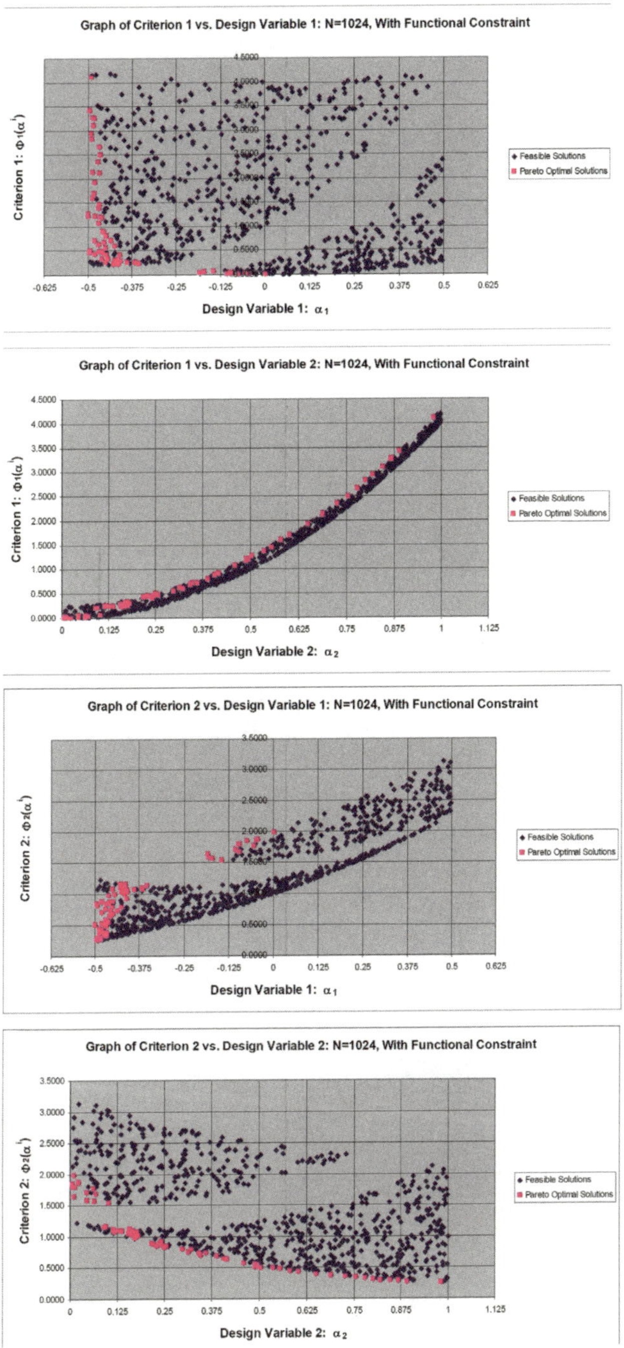

Figure A.10 Graphs of criteria versus design variables for $N = 1{,}024$ trials, with a functional constraint.

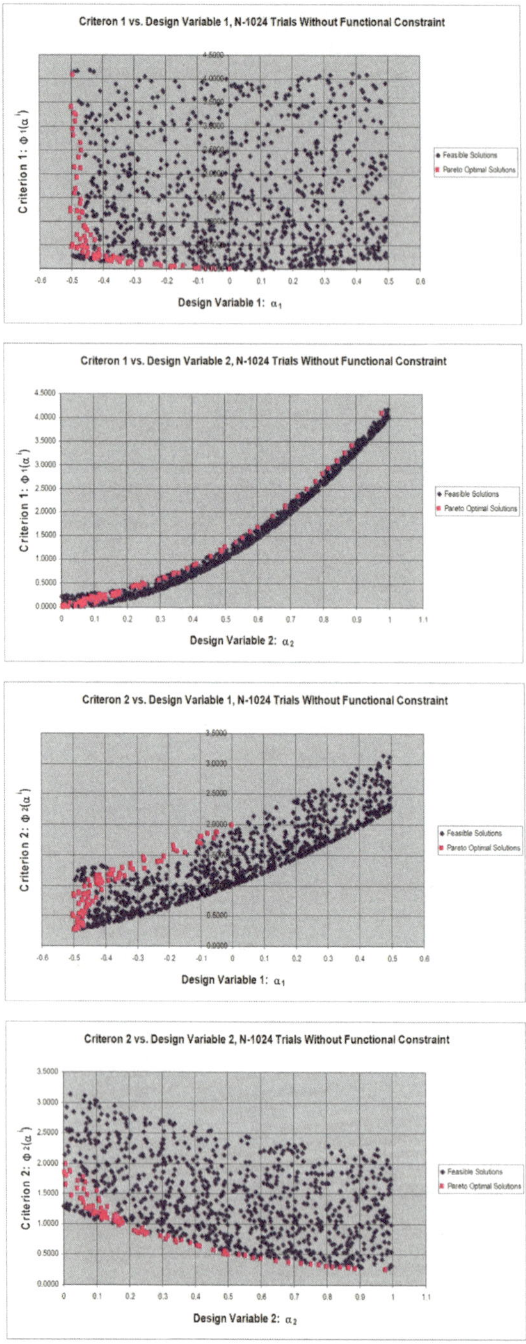

Figure A.11 Graphs of criteria versus design variables for N= 1,024 trials, without a functional constraint.

References

[1] Sobol', I. M., and R. B. Statnikov, *Selecting Optimal Parameters in Multicriteria Problems*, 2nd ed., (in Russian), Moscow: Drofa, 2006.

[2] Bartel D.L., and R. W. Marks, "The optimal design of mechanical systems with competing design objectives," *Journal of Engineering for Industry, Trans. ASME,* Vol. 96, No. 1, 1974, pp.171–178.

About the Authors

Roman Statnikov is a professor and principal research scientist in the Russian Academy of Sciences, Mechanical Engineering Institute, Russian Academy of Sciences, and since 2002 he has been a senior researcher and instructor at the Naval Postgraduate School in Monterey, California. Professor Statnikov has developed the parameter space investigation (PSI) method for the correct statement and solution of multicriteria problems jointly with Professor I. M. Sobol'. The PSI method has been widely integrated into various fields of industry, science, and technology. The PSI method has been used in design of the space shuttle, nuclear reactors, missiles, automobiles, ships, aircraft, metal tools, and bridges. Professor Statnikov has developed an interdisciplinary course "Multicriteria Analysis in Engineering" and taught it online since 2004 at the Naval Postgraduate School. He is also a consultant for many companies.

Alexander Statnikov is an assistant professor in the Center for Health Informatics and Bioinformatics and Department of Medicine, New York University School of Medicine, New York, New York. He is also the director of the Computational Causal Discovery Laboratory.

Index